LOVELAND PUBLIC LIBRARY
000591439

5/24/18
$ 14.95
AS-14
5/18

WHAT THE FUTURE LOOKS LIKE

D0089794

Withdrawn

WHAT THE FUTURE LOOKS LIKE

Scientists Predict the Next Great Discoveries and Reveal How Today's Breakthroughs Are Already Shaping Our World

Edited by
JIM AL-KHALILI

THE EXPERIMENT
NEW YORK

What The Future Looks Like: *Scientists Predict the Next Great Discoveries and Reveal How Today's Breakthroughs Are Already Shaping Our World*
Selection, introduction, and Chapter 18 ("Teleportation and Time Travel") copyright © 2017, 2018 by Jim Al-Khalili

Other chapters copyright of the author in each case © 2017, 2018 by Philip Ball, Margaret A. Boden, Naomi Climer, Lewis Dartnell, Jeff Hardy, Winfried K. Hensinger, Adam Kucharski, John Miles, Anna Ploszajski, Aarathi Prasad, Louisa Preston, Adam Rutherford, Noel Sharkey, Julia Slingo, Gaia Vince, Mark Walker, Alan Woodward

Originally published in the UK as What's Next? edited by Jim Al-Khalili © 2017.

First published in North America by The Experiment, LLC, in 2018.

All rights reserved. Except for brief passages quoted in newspaper, magazine, radio, television, or online reviews, no portion of this book may be reproduced, distributed, or transmitted in any form or by any means, electronic or mechanical, including photocopying, recording, or information storage or retrieval system, without the prior written permission of the publisher.

The Experiment, LLC | 220 East 23rd Street, Suite 600 | New York, NY 10010-4658 | theexperimentpublishing.com

Many of the designations used by manufacturers and sellers to distinguish their products are claimed as trademarks. Where those designations appear in this book and The Experiment was aware of a trademark claim, the designations have been capitalized.

The Experiment's books are available at special discounts when purchased in bulk for premiums and sales promotions as well as for fund-raising or educational use. For details, contact us at info@theexperimentpublishing.com.

Library of Congress Cataloging-in-Publication Data

Names: Al-Khalili, Jim, 1962- editor.
Title: What the future looks like : leading science experts reveal the surprising discoveries and ingenious solutions that are shaping our world / edited by Jim Al-Khalili.
Other titles: What's next? (London, England)
Description: New York : The Experiment, [2018] | Originally published as: What's next? / edited by Jim Al-Khalili (London : Profile Books, 2017). | Includes bibliographical references and index.
Identifiers: LCCN 2017052652 (print) | LCCN 2017060642 (ebook) | ISBN 9781615194711 (ebook) | ISBN 9781615194704 (pbk.)
Subjects: LCSH: Science—Forecasting. | Technological forecasting.

Classification: LCC Q175 (ebook) | LCC Q175 .W546 2018 (print) | DDC 501—dc23

LC record available at https://lccn.loc.gov/2017052652

ISBN 978-1-61519-470-4
Ebook ISBN 978-1-61519-471-1

Cover and text design by Sarah Smith

Manufactured in the United States of America

First printing April 2018

10 9 8 7 6 5 4 3 2 1

CONTENTS

MAKING THE FUTURE

Engineering, transportation, and energy

THE FAR FUTURE

Time travel, the apocalypse, and living in space

Introduction

Jim Al-Khalili

According to Einstein's theory of relativity, the future is out there, waiting for us—all times, past, present, and future, preexisting and permanent in a static four-dimensional space-time. And yet our consciousness is stuck in an ever-changing now, crawling along the time axis, welcoming the future as we gobble it up, then leaving it in our wake as it transforms into the past. But we are never able to see what is ahead of us. It is an incontestable fact that we cannot predict the future, despite the claims of psychics and fortune-tellers.

On a metaphysical level, whether our future is predestined or open, whether our fate is sealed in a deterministic universe or whether we have the freedom to shape it as we wish, is still a matter of debate among scientists and philosophers. Sometimes, of course, we can be reasonably confident what will happen—indeed some future events are inevitable: the sun will continue to shine (for another few billion years anyway), the earth will continue to spin on its axis, we will all grow older, and the English soccer team I follow, Leeds United, will always leave me disappointed at the end of every season.

In other ways, the future can unfold in completely unexpected ways. Human culture is so rich and varied that very often events happen in ways no one could have predicted. So, while there will have been a few who foretold Donald Trump's US election victory in 2016, no one can (thus far) predict when and where the next big natural disaster—maybe an earthquake or a flood—might strike.

Predictions about the way in which our lives will change thanks to advances in science and technology are spread across that wide expanse between the inevitable and the utterly unforeseen. The most reliable, and imaginative, soothsayers when it comes to conjuring up the future are usually science fiction writers, but how many of them before 1990 described a world in which the internet would connect all our lives in the way it does today? The World Wide Web still sounds fantastical when you stop to think about it.

So how does one compile a book, in the second decade of the twenty-first century, on what scientific advances are awaiting us, whether they are just around the corner, five years or ten years from now, or they are further off in the future, way beyond our lifetimes?

Some of the essays in this book serve as dire warnings about the way our world will be shaped, whether by nature or human activity, if we don't do something now. Solutions to our global problems will require financial, geopolitical, and cultural elements as well as scientific and engineering ones, but it is clear that harnessing our knowledge of the natural world, as well as the use of innovation and creativity in the technologies that exploit any new science, is going to be more vital than ever in the coming decades. So these essays are also beacons of hope, because they show how science can mitigate worst-case scenarios, such as the damaging effects of climate change, overpopulation, or the spread of pandemics through microbial resistance.

It is also undeniably true that the implementation of new technologies, whether in AI, robotics, genetics, geoengineering, or nanotechnology, to name but a few exciting current areas of rapid advancement, must be carefully considered and debated. We cannot afford to allow ourselves to be propelled headlong into an unknown future without carefully exploring the implications, both ethical and practical, of our discoveries and their applications. Many examples come to mind, such as the way robots are already beginning to replace humans in the workplace, how we can best guard against cyber terrorism, or the way we use up our natural resources while destroying habitats and threatening the ecosystem as the world's population grows both in size and greed. But I am painting a bleak picture, and our future need not look like that.

It is important to remember that scientific knowledge in itself is neither good nor evil—it's the way we use it that matters. You can be sure that within a decade or two we will have AI-controlled smart cities, driverless cars, augmented reality, genetically modified food, new and more efficient forms of energy, smart materials, and a myriad of gadgets and appliances all networked and talking to each other. It will be a world almost unrecognizable from today's, just as today's world would appear to someone in the 1970s and 1980s. One thing we *can* say with certainty is that our lives will continue to be completely transformed by advances in our understanding of how the world works and how we harness it.

Some of the contributions in this book paint a relatively reliable picture of the future. This is because the science they describe is with us in embryonic form already and we can see clearly how it will mature in the years to come. Others offer more than one scenario describing how our future will unfold, not because we don't understand the science or because its application may throw up surprises, but because the path we take depends on how that scientific knowledge is used. These will be decisions we must make collectively as a society and that require responsible politicians as well as a scientifically literate populace.

Inevitably, certain topics, whether driverless cars, genetic engineering, or the so-called Internet of Things, are covered by more than one contributor. This is deliberate, since it gives the reader more than one perspective on the different ways all our lives will change in the coming decades. It also highlights that many of these new technologies are interconnected and act to drive each other forward.

Some of the essays, particularly those toward the end of the book, are inevitably more speculative as we peer further ahead. In my own contribution at the very end of the book, for example, I take a look at the very distant future, way beyond our lifetimes. But then where would any self-respecting book on the future of science be without some mention of teleportation and time travel?

It is fair to say that some of the early chapters in the book strike a rather somber note in describing future scenarios, particularly if current warnings are not heeded, while others hail the wondrous technological advances just around the corner that will enrich our lives. An important point to make at the outset, however, is that this collection of essays is meant to be neither celebratory nor alarmist, but aims instead to paint as honest and objective a picture as possible, as seen by the world's leading experts in their fields, of what our future will look like. The one thing all the essays have in common is this: They are all based on our current understanding of the laws of nature—science fact, not science fiction. They do not predict a fanciful or far-fetched future and they do not appeal to magic or fantasy. I like to think that their sober assessment—their very groundedness—makes for a far more honest—and, yes, exciting—read.

THE FUTURE OF OUR PLANET

Demographics, conservation, and climate change

Loveland Public Library
Loveland, CO

1

Demographics

Philip Ball

The world changes because we do. Like most frequently overlooked truths, it's obvious once you say it. The future will be different not simply because we will have invented new technologies but because we will have chosen which ones to invent and which ones to use—and which, thereby, we will permit to change us. Some of these technologies will surely solve a few long-standing problems as well as create some new ones; some will touch hardly at all the big challenges that the future threatens. At any rate, we won't foresee the future simply by placing our present-day selves amid extrapolated versions of our natural and artificial environments. So: How will we live differently—and how differently will we live?

Mouths to feed

One of the biggest drivers of change today is population growth, which is possible only because of technological change. We could not sustain a planet of 7½ billion without the changes in agriculture and food production that have taken place since the nineteenth century—in particular, the so-called Green Revolution that, in the middle decades of the twentieth century, combined the development of high-yielding crop strains with the availability of artificial fertilizers. Without those advances, billions would probably have starved.

But it is not clear that we can sustain a planet with more than 9 billion people on it, as is predicted for 2050, without substantial further innovations, particularly in food growth and production and water resources. Most of the population growth will be in Africa and Asia—in countries that lack economic and infrastructural resources to easily accommodate it.

There is no guarantee that agricultural productivity is going to increase in line with population. Climate change—which can increase soil erosion, desertification, and loss of biodiversity—is expected to decrease productivity in much of the world, including many of those regions where population growth will increase demand for food. Such changes are now coupled to the vicissitudes of the market through globalization; changing demand or priorities in one place (such as the cultivation of crops for biofuels) can have a significant impact on food production or provision elsewhere. This means that food security is going to remain high on the agenda of concerns about a sustainable future for the world. Already, a food price spike in 2008 triggered widespread social unrest and led to the fall of the government in Haiti, while increases in food prices in 2011 have been implicated in the "Arab Spring" uprisings in north Africa.

The outlook is no better for water. Three quarters of a billion people currently face water scarcity. This figure could rise to 3 billion by 2025, while freshwater reservoirs are already oversubscribed in arid regions ranging from the American Midwest to the North China Plain.

You could see all of this as a tale of woe, a forecast of disaster, civilizational breakdown and the end of days. Or you could regard it as a to-do list for the political and technological challenges ahead. But perhaps more than anything else, it's a reminder of what matters in the future. Yes, personalized medicine and intelligent robots, asteroid mining and organ regeneration all sound very thrilling (or satisfyingly chilling, depending on your view)—and maybe they will be. But the age-old problems of humankind—How will we feed ourselves, and what will we drink?—will not be going away any time soon. Indeed, it may be these issues, more than any technological innovations in information, transportation, or medicine, that will dictate our patterns of personal and international interaction.

What we need, then, is a framework of sustainability. The word is used often enough without having really thought through what it would actually take to bring it about, or what it would look like once achieved. Some economists discount warnings of unmanageable population growth as alarmist, figuring that human innovation and ingenuity will sustain us much as they ever have. Others point out that the economic imperative for open-ended growth, driven by market forces and relegating undesirable issues such as pollution to the sidelines, is deluded and impossible in the long term. Both sides of the debate can marshal data, or at least narratives, to fit their view, but what tends to get overlooked is that science already has a framework—thermodynamics—that places strong restrictions on the options. Nothing happens—not food production, not the appearance of new ideas, not the metabolic maintenance of a society—without a cost in energy and the consequent generation of waste. To put it simply, there are no free lunches. Societies are complex ecosystems, but they are ecosystems like any other: webs of interaction, requiring energy, fighting entropic decay, adaptive but also vulnerable to fragilities. Creating a true science of sustainability is arguably the most important objective for the coming century; without it, not an awful lot else matters. There is nothing inevitable about our presence in the universe.

The changing face of us

Who, though, will "we" be?

A combination of increasing longevity and decreasing birth rates means that the population globally is becoming older on average. By 2050, the US population aged sixty-five and older is projected to more than double to around one in five Americans, and a third of the people in developing countries will be over sixty, which will, among other things, place greater strain on healthcare requirements and shift the proportion of the working population.

We must also ask "*where* will we be?" In the early twenty-first century the world population passed a significant marker when a 2007 United Nations report announced that more than half the people on the planet now live in cities. For most of humankind the future is an urban one.

There are now many megacities with populations of over ten million, most of which are in developing countries in Asia, Africa, and South America: Mumbai, Lagos, São Paulo, and Manila are examples. Nearly all the population growth forecast for the next two decades will be based in such cities, especially in developing countries, and by 2035 around 60 percent of the world's population will live in urban areas.

In the old tales, you set out into the wide world to seek your fortune; today you look for it in the city. Many people come to cities from the surrounding countryside in the hope of a better life, but they don't necessarily find it. Many cities can't cope with such an intake: for example, 150 million city dwellers now live with water shortages. In addition, many fast-growing cities in low-lying coastal areas will be at increasing risk of flooding as sea levels rise and extreme weather events become more common, as is forecast by climate change models.

You don't need a crystal ball to predict a continuation of the waning global influence of the United States, nor to foresee the cloud hanging over the European unification project. But if you harbored any doubts, a glance at the changes in the world's largest cities tells us something about where the action is likely to be in the years to come. In 1950 these were, in order of size: New York, Tokyo,

London, Osaka, and Paris. In 2010 the top five had become: Tokyo, Delhi, Mexico City, Shanghai, and São Paulo. By 2030 the list is predicted to read: Tokyo, Delhi, Shanghai, Mumbai, and Beijing. To find the epicenters of the future, go east.

Of course, it is one thing for a city to be growing, quite another for it to be thriving, as is all too evident among the favelas of Rio de Janeiro and São Paulo. All the same, there seems little doubt based on current form that China and India will continue their growth into global superpowers. Over the next twenty years, China is expected to construct two or three hundred entirely new cities, many with populations exceeding a million. In fact, the equivalent of a city of about a million and a half people is added to the planet every *week*.

But what will a city of the future look like? An artist's impression, all gleaming glass and chrome topped with greenery, can be very enticing—but also misleading, because there is no single future of the city. Some look likely to become more user-friendly, more green and vibrant. Others will sprawl in a morass of slums, perhaps punctuated in the middle by a glittering financial district, with disparities in wealth that will dwarf those of today. Can we even plan a successful city, or must they always grow "organically," as influential urban theorists Lewis Mumford and Jane Jacobs have argued, if they are to be vibrant and thriving rather than soulless and sterile?

Some researchers believe we have little hope of answering such questions unless we develop a genuine "science of cities," rather than relying on the often arbitrary, prescriptive, and politicized dreams of urban planners and architects. There are a few glimmers of such a nascent discipline, not least in the realization that some things apply to pretty much every city, regardless of size and character. Cities depend on economies of scale: the bigger they are, the less they require per capita in terms of infrastructure and energy use, the more average earnings increase, and the more they become innovation machines. But everything grows faster with scale, good and bad: bigger cities have higher rates of crime, theft, and infectious disease, and a faster pace of life generally, whether it is the rate at which businesses rise and fall or even the speed at which people walk. It seems you can't have the advantages of cities without their drawbacks. So take your pick, if you're lucky enough to have the choice.

The trend of migration to cities is a part of a much broader exodus across the planet. The United Nations estimates that currently over 200 million people have migrated from their home country to another, and around 740 million have relocated within their own country. Over the past several decades, a great deal of this movement involved migration from rural and mountainous areas to cities.

Why all this movement? In low-income countries most people relocate for economic reasons, looking for better employment opportunities, higher wages, or diversification of livelihood, especially if agriculture becomes unsustainable as a way of life. Some seek education, or move to be with other family members. Some wish to escape political or cultural persecution, war and conflict, as in Syria; some are forcibly relocated for sociopolitical reasons, as in the construction of Chinese dams. And some are compelled to flee from environmental hazards, such as floods, infertile agricultural land, or lack of water resources.

Climate change will inevitably increase migration in the years and decades to come, but that doesn't mean it is meaningful to speak of "climate migrants." Environmental change can interact with other drivers of migration in complex ways. Rural drought, added to the economic and political crisis in Zimbabwe, has led 1.5–2 million people to flee to an often hostile reception in South Africa since the turn of the millennium. What's more, movements prompted by environmental change can blur the distinction often made in policy and legal circles between "migration" and "displacement";

the former is considered a choice, the latter an enforced necessity. It is not always clear when local conditions can be considered to have deteriorated sufficiently for migration to be involuntary rather than voluntary. At any rate, there can scarcely now be any question after the recent European experience that migration and immigration will be dominant political themes for years to come.

Identity technologies

All this burgeoning change might seem a far cry from life in rural Africa or amid the nomads of Mongolia—except perhaps for one factor. Thanks to telephone networks, they're linked in.

Two out of every three people in the world now own a mobile phone (or at least, a phone plan), even in less developed countries in sub-Saharan Africa. These devices are mainly how we (tele)communicate today. Internet access doesn't yet show quite the same spread: In developed countries, four out of five households have it, but this drops to well below one in ten for the least developed countries. Fears of a technological or digital divide are warranted, but it's not a simple equation. That divide is, not surprisingly, also evident in the age demographic: in the US, more than 99 percent of people polled in 2016 between the ages of 16 and 24 said they had used the internet in the last three months, but only 39 percent of people over 75 had done so. And a recent poll in the US showed that while 41 percent of Americans aged sixty-five and older don't use the internet, only 1 percent of those between eighteen and twenty-nine stayed offline.

Access tells only part of the story; mobile networks have also shifted patterns of use toward an "always on" mentality. The so-called Generation Z, who were born in the 1990s and have never known a time that lacked these facilities, is now reaching adulthood, and a 2011 survey of Britons aged 16 to 24 found that 45 percent felt happiest when they were online. Many businesses now expect employees to be constantly contactable by mobile phone and email; equally, domestic and personal matters can be managed from the work desk, breaking down the barrier between separate identities of work and home.

There are plenty of statistics like this, but exactly what they mean isn't obvious. A straightforward extrapolation of current trends implies that three quarters of the world will have mobile phones by the end of the next decade. But the consequences of phone access for a farmer in Kenya or a Mongolian nomad are very different from those for a city trader in London.

The spread of information technologies and social media justifies their description as "transformative" and indeed "disruptive" technologies—but what will they transform and disrupt? It was exhilarating to imagine that the Arab Spring uprisings of 2011 were "Twitter revolutions," but the evidence for that has largely evaporated, and in any event that idea told us nothing about how they would unfold subsequently.

An increasingly "info-connected" world is just one facet of a trend toward greater interdependence affecting, and affected by, trade, travel, disease, censorship, privacy, and a great deal else. A heady stew, in other words, and no one can know how it will taste. Here are a few suggestions of what experience so far can teach us:

- Interconnectivity doesn't mean inclusivity. On the contrary, it may produce a Balkanization of views that coarsens political discourse and supports or hardens extremist views. There is little sign that the internet or social media encourages broadmindedness and debate; in some ways they are set up to insulate us from dissent or challenge, for example, by offering to personalize news feeds. It used to take some effort to find Holocaust-denying pseudohistory; now it's one click away.

- Just as information technologies may serve to amplify existing prejudices and misconceptions, so they amplify inequality. In business and trade, in arts and entertainment and fame, markets have become ever more inclined toward "winner takes all." This, psychological studies show, is precisely what to expect from rating systems in which you can easily see what choices others are making.

- If a job can be done by a robot, it probably will be. Already there is a significant sector of the financial market in which all activity is carried out by automated trading algorithms. It happens faster than humans could possibly manage, it has its own rules, and we don't yet really know what they are. This automation will expand into ever more sophisticated jobs, including in healthcare and education. There may be benefits, certainly: robot doctors never sleep, you might not need to wait weeks for an appointment, and the robot might know more about your health (from implanted monitors and genomic data) than a human doctor ever did. But automation will transform labor markets—and a clear lesson from history is that those who have no stake in a society's productivity, rather than enjoying greater leisure, are deprived of economic power.

- Your greatest asset might not be your skills, knowledge, or even wealth, but your reputation—how you are rated by others on online forums, say. That means you will need to manage and curate that reputation well—or perhaps to employ someone else to do so, as companies already do.

These trends don't clearly point in any single direction, and indeed many contain internal contradictions: lies are easier to expose, but are easier to spread, too. Most important, none of these changes is happening in a sociopolitical bubble: What they mean in China is not the same as what they mean in Sweden or Iran.

It's safe, though, to draw one implication for our future selves: Identity is far less fixed, and far more multifaceted, than it used to be—or at least, than it was thought to be. We have multiple identities that surface in different situations, often overlapping and increasingly blurred but defining our views and choices in distinct ways. In particular, traditional social categories that defined identity, such as age, class, and nationality, are becoming less significant, as are distinctions between public and private identity. Old definitions of identity based on class, ethnicity, and political affiliation may cede to new divisions, marked, for example, by distinctions of urban/rural or well/poorly educated.

If traditional attributes of individual identities become more fragmented over the coming decade, communities might be expected to become less cohesive. The result could be reduced social mobility and marginalization, creating dangers of segregation and extremism. On the other hand, hyperconnectivity can also produce or strengthen group identities in positive ways, offering new opportunities for community building. Will our increasingly connected lives and identities work for better or worse? Well, both—and ever more so.

The future of democracy and religion

Francis Fukuyama's 1992 book *The End of History and the Last Man* has become a favorite whipping boy of futurology: Look now and laugh at the idea that, after the fall of the Berlin Wall and the Soviet Union, liberal democracies are the logical final expression of every developed state. But there is more reason than ever now to doubt Fukuyama's cozy forecast. Not only is it clear that stable democracies remain as stubbornly elusive as ever in large parts of the world—and certainly don't arise as if by magic from the overthrow of a dictatorship—but it also can't be taken for granted that, once they have arrived, they are here to stay. Demagogic populism in Europe and the United States is, at the time of writing, threatening to transform liberal democracies into the "Strong Man" regimes more often associated with Russia, China, and southeast Asia, sustained by coercion, corruption, and collusion. There is serious discussion about whether the terms "liberal" and "democracy" can remain partnered indefinitely, and whether unrestrained capitalism—with its economic fictions and its tendency to promote inequality and resentment—can be consistent with either or both.

In short, Western commentators are no longer quite so sure that they have mastered the best form of governance after all, much less that they have any notion of how to foster it elsewhere. According to political scientist David Runciman, the advantage of democracy—that it can bounce back from all kinds of shocks—is also its Achilles' heel, since it lacks any incentive to truly learn from the past. Muddling through seems a good enough policy, until it isn't.

The one thing it seems safe to say is that we should stop thinking of politics as some kind of chemical reaction that fizzes for a while before reaching a static, unchanging equilibrium. Change seems to be the only certainty—and increasingly political scientists speak of it as a "discontinuous" process, happening not gradually but in sudden, seismic shocks.

Among the changes we might anticipate are those in the status of religion. The question is perhaps not so much "whither religion" but "which religion"? The "big four"—Islam, Christianity, Buddhism, and Hinduism—seem increasingly to be eclipsing others. Atheism, professed by around 16 percent of the world's population, although much higher in Western Europe, is spreading more slowly than the major religious faiths (except for Buddhism, which is actually declining proportionately). The global proportion of Muslims is growing fastest, and is expected to equal that of Christians (around 30 percent) by 2050.

However you feel about those trends, it makes sense to regard them in much the same spirit as we study the spread of other cultural traits, such as language—especially as, like them, religion cannot be divorced from other factors, such as population growth and economic development. And like those factors, religious belief will continue to shape lives in profound ways, for better or worse. History makes it clear that religion need not be anti-intellectual, anti-science, anti-democratic, anti-humanist. But it also shows that it can be.

In the long run...

The best science fiction is never about predicting the future. Those stories from the genre that find real traction—think of *The War of the Worlds*, *Brave New World*, *Nineteen Eighty-Four*, *Blade Runner* (and the Philip K. Dick novel on which it was based), or *Gattaca*, the 1997 movie about genetic segregation—use the inventive freedom the future offers to explore the anxieties of the present. So when that promised future of jet packs, moon bases, and robot servants fails to materialize, we have no cause to complain. That was never the point.

Yet even the best science fiction sometimes makes the error of imagining what technology will do to us, without recognizing how much technology responds to us. Technologies are rarely, if ever, truly foisted upon us, however much that might feel to be the case. They arrive because, as a society, we accept, welcome, and ultimately normalize them, often to the extent that they become more or less compulsory. Before mobile phones and social media, we never really recognized how narcissistic we are, how desperate we are to escape the reality of our surroundings and to assuage a sense of loneliness. We didn't appreciate quite how much societies run on trust (e-commerce), how averse we are to conflicting views (the echo chamber), how fascinating we find the mundane (reality TV), how nasty anonymity can make us (trolling).

So futurology can and must force us to hold up a mirror to ourselves. It is one thing to imagine, say, a transhumanist future in which we are immortalized by melding our minds and bodies with information technologies, downloading our thoughts to some quantum hard drive. That might or might not be sheer fantasy (I happen to think it is)—but the fantasy is revealing. It suggests that our relationship with death and dying will be a driver of social change, one way or another. Likewise, I'd suggest that many of the forecasts in this book are better viewed not as what the future will be, but as what we yearn for it to become.

When we ask of the future, "Who will we be?," government departments tend to produce sober reports replete with maps and statistics, mostly based on gradual extrapolation from the here and now. Futurologists, on the other hand, imagine "discontinuous change" and "singularities": an abrupt shattering of the status quo, a fracture of the graph created by an unforeseen technology or a political crisis. Artists and writers, meanwhile, offer flights of imagination—some wild, even lurid, often tinged with satire. We will live in a Brave New World, a race grown in vats of a Central Hatchery; or on a planet that has vanished beneath a thicket of megatowers, each one a multistory nation; or among the ruins of Armageddon, telling legends of technological wonders in a debased Chaucerian vernacular. We need all these numbers and transitions and visions, not because they will give us the answer, but because we must try to spot the traps we set for ourselves. As the American writer Richard Powers has put it: "People want everything. That's their problem."

2

The biosphere

Gaia Vince

Dawn on Costa Rica's Pacific coast and the blue-black silhouette of a man at the water's edge gradually enlivens under a pinkening sky. I switch off my flashlight. Jairo Quiros Rosales and I are the only people to be seen on this broad black beach, whose volcanic sands stretch north for several miles. Jairo is beckoning me, so I hurry down to him, scanning the beach and murky shoreline. As the light grows, I make out the funereal vultures flecking the distance, and assorted mutts appear from the gloom to sniff the night from the sands.

And then I see them: about 100 yards farther up the beach, like strange, regularly humped stones, hundreds of olive ridley sea turtles are making their way from the ocean onto the beach to lay their eggs. This is the *arribada*—"the arrival" in Spanish—and I have been waiting more than a month to see it. In my eagerness, I grab Jairo by the arm and hustle him faster toward the turtles. He's a little surprised, but smiles. Although we haven't met before, we've built up a certain familiarity over the phone—or, rather, my continual nagging phone calls asking when the *arribada* will happen have made me fond of the shy Costa Rican researcher. We speak in Spanglish—his English is better than my Spanish, but as with most conversations, there are concepts that work best in different languages. *Arribada* is one of them.

Most marine turtles nest individually at various times during the year so that their young hatch at unpredictable times and places to avoid predators. But olive ridley (and the closely related Kemp's ridley) turtles evolved a unique, mass-nesting strategy. By synchronizing their egg-laying, so many hatchlings are produced at the same time that the predators cannot consume them all and are overwhelmed. It's known as "predator swamping." The mass emergence of olive ridley turtles happens a few times a year in just a few places around the world, and Ostional beach in Costa Rica is one of them.

As we walk in symphony along the glossy shoreline, turtles stream out of the sea like tanks invading the beach—a maternal armada of ancient reptiles driven forward by hormonal compulsion to deposit their precious cargo. Jairo points out to sea where a line of carapaces bob parallel to the shore, little heads poking up periodically to breathe, waiting their turn. Ahead of us the beach begins to undulate with heart-shaped, olive shells crawling over and past each other in urgency. There are perhaps tens of thousands now crowding the beach. Some, having done their business, are on their way back to the ocean, heaving their heavy shells against the tide of pregnant comrades on flippers poorly suited to terrestrial marches. Spent, they wait at the shore for incoming waves to sweep them out to sea.

I'm absorbed in the wonder of it all. "*Maravilloso*," Jairo agrees, looking at my face. He's been up since 2 AM and seen this many times, and yet he, too, is visibly moved. Marine turtles are usually hidden from us in their underwater world, while we are bound to dry land. I've seen them close up, rarely, while diving, where they move effortlessly and with surprising grace. It is unusual to see large wild animals up close, and to see so many—to be surrounded by so many—is incredible. Olive ridley turtles, like all marine turtles, are threatened with extinction because of us.

These hard-shelled, living dinosaurs are just one of 23,000 species around the world that are currently threatened with extinction. Humans now dominate the planet so extensively that we are pushing wild animals and plants off the face of the Earth. We already use more than half of the world's land for our food, cities, roads, and mining; we use more than 40 percent of the planet's net primary productivity (that's everything produced by plants and animals); and we control three quarters of all freshwater. We are now the most numerous big animal on Earth and second on that list are the animals we've created through breeding to feed and serve us. Our planetary changes now threaten one in five species with extinction, roughly one thousand times the natural rate of extinction—we have lost half of our wildlife in the past forty years alone, and biologists warn that we are entering the sixth mass extinction event in Earth's history. To put this into perspective, previous such events, including the one that killed off the dinosaurs, were the result of cataclysmic activities such as giant asteroid impacts or supervolcanic eruptions.

The natural world is reeling from our global impacts, and for turtles, like so many other species, it's only getting worse. Should we care? What does it really matter if we lose a bunch of animals we rarely even see? Humanity's relationship with the natural world is a complicated one—to understand our current extinction pattern, we need to look at the way our human lives and livelihoods, as well as our desires and motivations, are enmeshed within our complex global environment.

For me, the story of the *arribada* offers a unique insight into a planetary problem that is far bigger than any individual players. The details are, of course, different for each animal or plant struggling to survive on a human-dominated planet, but the human emotions and drivers are universal. What makes Ostional so extraordinary is that local residents have found a way to make use of their natural resource but also to protect it. And that is the key: We cannot protect the world's wildlife unless we also protect the needs of the humans that rely on it.

As we walk along the beach, Jairo is counting the turtles, estimating numbers between regularly placed ranger posts by noting shells and the curious tank tracks they leave in the sand on their commute up and back. Later, researchers will make a more accurate count, he says, but on first reckoning, there are more than ten thousand turtles so far in the *arribada*. Some of them will be tagged and measured and logged on an international database so that their movements can be tracked. First, he wants to show me something.

"*Venga* [come on]," he urges, and leads me up the beach to where the damp flatness eases to soft dry dunes. Here, above the tideline, is where the turtles nest. A new arrival has made her way up here and is beginning to dig at her chosen spot. Jairo and I squat down to watch her. With her front flippers she spades the sand, flicking it left and right, frequently covering my feet. On and on she goes, digging a broader and deeper hole, continuing a ritual that has been practiced since the time of the dinosaurs—a time ruled by reptiles, when the planet was warmer and wilder, animals grew to an enormous size (like the ten-foot-long Archelon turtle of the Cretaceous or the more recent two-ton Stupendemys turtle), a shell was adequate protection against predators, and there were no humans or anything resembling us on Earth. In just a few minutes the hole is ready for her to gently reverse into, lowering herself tail first. From here, she uses her back flippers to scoop out a deeper depression, flicking more sand over us in a manner I can only interpret as deliberate because she is looking directly at me while she does it.

When the hole is finally to her satisfaction, she readies herself for egg-laying. Here, her labors begin. Her shell heaves with obvious effort and her eyes stare unresponsively as she enters a trancelike state. Beneath her, in the carefully prepared nest that she has judged to be of the right temperature, depth, and distance from the ocean, one by one she is depositing her eggs—her evolutionary raison d'être, the genetic material that links her to her mother, grandmother, all the way back to the Cretaceous and, in some fundamental way, to her contemporary distant cousin: me. Her breathing comes strongly and in her efforts moisture gathers from her nostrils. Across the divide of animal class, from mammal to reptile, I feel great empathy for this mother. Around us, in this vast maternity ward, other mothers are flicking sand or laying eggs. And between them stalk the vultures and dogs, biding their time, waiting to dig up the newly laid eggs.

Each turtle lays around one hundred eggs, but of more than ten million eggs laid in an *arribada*, only around 0.2 percent typically survive to hatch. And of the hatchlings, just 1 percent are thought to make it to adulthood. Part of the problem is the *arribada* itself, which lasts for around five nights. So many turtles laying on a comparatively small stretch of beach means that turtles arriving on subsequent nights dig up and damage the previous night's eggs, causing bacterial infections to destroy both sets of eggs. And, because the incubation period is at least forty-five days, but *arribadas* usually occur at monthly intervals, a nesting turtle may dig up and ruin eggs from a previous *arribada*, too.

Vultures are already feasting on the torn remains of scattered eggs. They cannot dig them up from the nests, but dogs can, and wherever there are humans, we bring dogs. The dogs here are a menace to the arriving turtles as well as to their eggs and hatchlings. Jairo shoos them away but they return quickly. He asks me if I want to see further, and I nod. Carefully, he brushes a bridge of sand away from the tail of our laboring turtle and I peek through. Down in her nest is a glistening clutch of white eggs, the size of ping-pong balls. From behind her tail, her fleshy ovipositor hangs down, and while I watch, mesmerized, it releases another precious ball, followed by a squirt of clear viscous protective fluid to coat the permeable eggs.

We watch a few more eggs drop down and then Jairo replaces the sand seal and we sit back.

Turtles are exquisitely adapted to their environment—they have survived almost unchanged since the Triassic—and they can live for more than a century in the wild, reproducing well into old age. But, in the Anthropocene, this age dominated by humans, they face perhaps their toughest challenge for more than a million years. Beaches where they nest have been disturbed by development and the sheer numbers of people and dogs. Our lights alone cause problems, confusing turtles and hatchlings that rely on moonlight to navigate. They are being killed and injured by boat impacts, entangled in fishing nets, and dying from ingesting plastic and other pollutants. Overfishing and the destruction of coral reefs, where they graze, is threatening their food supply. And climate change is also hitting them—rising sea levels and associated beach erosion reduce the area available for nesting and some beaches have become unusable for them, and warmer temperatures are causing worrying sex changes. The sex of a turtle is dependent on the temperature of the incubating egg. Warmer eggs develop into females, cooler ones into males. Biologists are reporting that global warming is already resulting in an imbalance in the sexes for a number of reptiles, with worrying consequences for mating and the survival of turtles. The previous month, very unusually, there was no *arribada* here, and a lack of available males is one of the reasons suspected.

However, by far the biggest threat to turtles is poaching. Around the world nesting female olive ridleys are slaughtered on the beach for their meat, skins, and shells, and their eggs are traded as a valuable delicacy. In the past twenty years, just one generation, the global population has been slashed by one third. The global illegal trade in the world's wildlife is worth more than $17 billion a year and threatens the stability of governments as well as human health—some 70 percent of infectious diseases have zoonotic origins. Illegal wildlife trade is often conducted by well-organized criminal networks that undermine governments' efforts to halt other illegal trades, such as arms and drug trafficking, and help finance regional conflicts. Traders use euphemisms or describe products as "captive bred" to sell on global internet sites, such as eBay. More than half of all illegally traded wildlife ends up in China.

Conservationists like Jairo and a few government rangers patrol this beach during *arribadas*, but they are little match for determined poachers, who sell their eggs as aphrodisiacs on the black market. On the Caribbean side of Costa Rica, I've seen turtle eggs openly sold and eaten in bars and cafés. Moin beach, on that coast, is used by endangered leatherback turtles, and poaching is so rife that young conservationists—many of whom are volunteers—race to nesting sites to dig up the eggs and rebury them in secret, safer locations. The poachers, many of whom are also involved in drug crime, have turned violent, threatening and attacking the environmentalists. In May 2013 a young conservationist, Jairo Mora, was collecting turtle eggs for reburial when he was kidnapped by poachers and murdered. No one was jailed for his murder, and Mora joins a growing list of environmentalists killed for protecting wildlife in Costa Rica and beyond. In 2015, the most recent year for which records exist, 185 environmentalists were killed protecting natural resources globally. Only a tiny fraction of such deaths result in convictions.

"Do you worry about your own safety, when you're out here alone at night?" I ask Jairo. "No, the Caribbean is different," he says. And then he admits, "Sometimes." Since Mora's murder, most conservationists have been too scared to patrol Moin beach, and poaching is reported to continue uncontested. And yet, people like Jairo value the conservation of wildlife as something worth risking their own safety for.

It is fully light now, almost 6 AM. Jairo is tired but smiling. The beach we've had all to ourselves is about to be invaded. "Our" turtle has finished laying her eggs—she carefully covers them with sand and lumbers down to the shore, returning with the other mothers to the ocean, having each played their part in the continuation of their species. I feel ridiculously proprietorial about this beautiful place, having watched its sands fill with turtles and witnessed the private efforts of a mother birthing the next generation, as night turned to day. From the village end of the beach, I see a band of about forty people approaching, carrying large rice sacks and baskets.

We humans have always exploited our environment's resources for food, energy, and all our other needs. We're brilliant at it, and it's led us to being such a successful species that we live longer and better than ever before and now dominate the world. In the past, our activities led to a few local extinctions, but there are now more than 7 billion of us and we act on an industrial, global scale, threatening the very resources we rely on. Are we just another part of nature doing what nature does: reproducing to the limits of environmental capacity, after which we will suffer a population crash? Or are we the first species capable of self-determination, able to modulate our natural urges and manage our plundering of the natural world so we can maintain habitability into the future?

For the past few decades, the community here has been trying something unique—a controversial experiment in conservation that aims to maintain a sustainable turtle population while benefiting the impoverished local village. Ostional is the only place in the world where harvesting olive ridley turtle eggs is legal.

Malena Vega comes over to me, a warm smile creasing her round face. We've spoken a couple of times on the phone and she's kindly offered to include me in the activities today. She looks up at Jairo and he confirms that there were more than one thousand nesting turtles on the beach, the minimum number required for legal egg collection. With a friendly wave, he sets off back to the research station for a nap before his evening work, and at that moment a horn sounds and egg collection can begin.

In the late 1980s, representatives from the village approached biologists who were studying the *arribadas* to ask if something could be done to legalize egg collecting within sustainable parameters. They were concerned about the huge numbers of poachers who were descending on the village, stealing the eggs and intimidating locals. A plan was drawn up with the government, and the self-regulated women-run Ostional Development Association was established to allow certain families to harvest a limited number of eggs on the first three mornings of an *arribada*. (These eggs would be damaged by subsequent nestings anyway, and researchers calculate a 5 percent greater hatchlings rate following earlier egg removal.) As part of the agreement, the community cleans the beach and protects the turtles and their eggs from poachers and manages the enormous numbers of tourists that descend on Ostional during *arribadas*. The eggs harvested are licensed for sale at the same price as chicken eggs to deter the black market, and the proceeds used for community projects.

All around me, local men and women are dancing a tarantella on the sand, stamping gently in bare feet to find nests. Everyone is wearing some sort of turtle motif—a necklace or printed T-shirt. One by one, they drop and burrow. There are very few turtles on the beach now—the next wave won't come ashore until nightfall. Malena squats beside me, her hands moving rhythmically in the sand to uncover the denser layer below. My heart drops like a stone to my stomach as I realize this is her nest that I watched being so carefully prepared, filled, and covered.

"Come here," Malena calls. I join her and she grabs my hand and pushes it down into the hole. "Can you feel them?" she asks.

I root around a bit but am relieved to only feel sand. Malena puts her own hand in and expertly retrieves a couple of eggs, which she puts in her sack. "Try again," she tells me.

This time, I scoop my hand, imitating Malena's angle, and feel the eggs. I pull one out. Malena applauds me and repeatedly dives back into the hole, bringing out handfuls of eggs for the sack. She empties the nest, re-covers it, and moves a couple of yards away to start another one. I sit, holding the egg I retrieved in my hand. It is soft and warm and leathery, denting in my grasp. I have become the thief of my nightmares, plundering the nursery as soon as the mother has left. This is the antithesis of the environmentally responsible culture I have been brought up with—stealing eggs is bad enough, but stealing the eggs of a protected species is unforgiveable.

"You can eat it raw like that," Malena calls. And she demonstrates, tearing the shell and popping the contents in her mouth. Around me, in remarkably short time, sacks have been filled with eggs. It is already hot out here on the black sand and everybody wants to get back to the shade of the village. I help Malena tie her sack and together we trudge back over the sand to her house. En route, with frequent pauses—turtle eggs are unexpectedly heavy—Malena, who is president of the Ostional Development Association, tells me how the project has changed the community. Ostional is a small, poor, rural village wedged between two rivers, the mountains, and ocean. During the rainy season, when the rivers flood, the village is cut off completely and must survive on whatever food they store. Many people deserted Ostional for the cities, where there was work. Now, she says, the egg licensing gives people a living wage and has paid for training, maternity cover, and pensions. People are returning to the village and making lives for themselves here. "The turtles are our lifeblood," she says. "We love them. They mean everything to us here."

"But isn't this simply legalizing the poaching that you were doing before?" I ask her. "Before, this was a dangerous place," Malena says. "The beach was dirty and full of poachers from everywhere. The police came and there were gun battles. My grandmother got shot by mistake, and she died. After that, we said: No more! This is our village and these are our turtles." I am struck by the fierce determination of this woman—a grandmother herself now—and what she and her band of female neighbors have achieved.

We need the world's resources now more than ever—to develop economies of poor countries and to support our growing population. But we need to find a way of *sustainable* exploitation. What Malena and her neighbors are trying in Ostional is also being tried in rainforests for timber and in the oceans for fish. It's too early to tell whether we are truly able to limit our extraction of these endangered resources to levels that are sustainable. But the early signs show that where the long-term needs of communities living in these vulnerable environments are included, ecosystems are managed better, perhaps because our uniquely human altruistic tendency to care for other species and look after the home environment is invoked.

We have reached Malena's house now, a simple wooden dwelling subdivided with panels, where she lives with her daughter, granddaughter, and a few chickens. "You must try a tortilla." She smiles, listing numerous health-giving properties. As she heats a pan on the stove, chops coriander and onion, and whisks up a bowl of freshly gathered turtle eggs, I ponder how strange and random are our endowments of value on living things—and how deadly the consequences. Many environmentalists believe the legal egg collection here contributed to the murder of Jairo Mora by ensuring a market for turtle eggs. Yet I can see how the importance of the egg market to this Ostional economy is giving the turtles far greater protection.

In this new human age, we can no longer leave the natural world to manage itself under our onslaught—that way, only the weeds will survive. If we want to see wild animals, we have to actively protect them. On a global level, turtles are of little real use to us. Nevertheless, I would not like to live in a world where they no longer exist. As we try to negotiate a path between the competing demands of the human and natural worlds, Ostional shows us that it needn't be one or the other. In truth, to protect the wildlife, you must also protect human life.

The turtle omelet was delicious.

3

Climate change

Julia Slingo

Setting the scene

Climate change will be one of the defining challenges of the twenty-first century; how we respond will determine our future prosperity, health and well-being, and the sustainability of Earth's natural environment. In 2015 over 190 nations agreed to act to limit the increase in the Earth's surface temperature to less than 2°C (3.6°F), and preferably to 1.5°C (2.7°F) if at all possible. Future generations may well look back on 2015 as a turning point in the struggle to align policy with science. But how should we make sense of all this, and the daunting prospect of what our response to the challenge of climate change might mean for how we live, and where we live?

Let's start with the two words—climate and change. The word climate describes a long-term average—typically over thirty years or longer—of factors such as wind, temperature, and rainfall, which we might regard as stable. It's what we expect year by year and season by season. Many societies and economies are now finely tuned to the current state of the climate—India is a good example, where the regular return of the monsoon rains is essential for water, food, and energy security, and any delay or failure of the monsoon can have a huge impact on the country's economy. But according to the well-known saying (sometimes attributed to Mark Twain, but most likely coined by Andrew John Herbertson, a British geographer and professor at Oxford)—"Climate is what you expect, weather is what you get." So when we think about climate we also need to think about the range of weather that makes up our climate, and especially more extreme weather, how often it occurs, and what its impacts on us are likely to be. In fact, we expect the most profound impacts of climate change to be associated with weather extremes—windstorms, floods, wildfires, storm surges, and heat waves.

The word "change" implies something that is different from the norm—but how to define the norm? Our weather and climate vary continuously on all sorts of timescales from hours to decades, and that variability is part of defining the climate. One way to define change is when the climate we experience falls outside the margins that we, as a modern civilization, have become used to; this is essentially how scientists have arrived at the statement that evidence for climate change is "unequivocal."

It's clear, though, when we look back in history, that the Earth's climate has always changed, for example in and out of recent Ice Ages, to the deeper past when the Earth was warmer and carbon dioxide much more abundant than it is today. So why do we worry now?

There are three reasons that set current climate change apart from past changes. The first is the source of the change. Today our climate is changing because the concentrations of greenhouse gases in the atmosphere, especially carbon dioxide, are rising—and rising fast. We are taking ancient carbon that was bound up over millions of years as coal, oil, and gas, and releasing it into the atmosphere through combustion in the space of just a few decades, to satisfy our insatiable appetite for energy to support our industries and lifestyles. And we are changing the way that the Earth's biosphere takes up carbon and helps to reduce the effects of our emissions, by clearing forests and making the oceans more acidic. And because carbon dioxide is a greenhouse gas, this unnatural and rapid increase is causing the Earth's temperature to rise, with all the associated impacts on our weather and the natural environment, such as more extreme heat waves, droughts, and floods, and the loss of sea ice and glaciers. When we look at past climatic changes we can see that they were driven by very slow variations in the Earth's orbit around the sun. These affected first the temperature, which then drove a response by natural ecosystems that control the amount of carbon dioxide in the atmosphere. So, past changes were driven by very different forcing agents, and although we can learn from them, they are not analogues of what we are experiencing today.

The second difference is the pace of climate change. Carbon dioxide levels are now over 30 percent higher than they were just over one hundred years ago—and indeed over 30 percent higher than anything the Earth has experienced for at least 800,000 years. The Earth's average surface temperature has so far risen by 1°C (1.8°F) over the past century and is on course to reach a rise of 2°C (3.6°F) or even more by 2050 unless we take serious action. These numbers may sound small, but we should recall that the change in Earth's temperature coming out of the last Ice Age was only 5°C (9°F)! By rapidly releasing ancient carbon we are also effectively cutting many of the natural feedback loops that allow ecosystems to evolve through a period of climate change. We can see this happening already in the loss of natural habitats and in the changes to the seasons that affect the life cycle of many species. The pace of change is too fast for many plants and animals to migrate or adapt. It seems likely that this disruption of our climate could have far-reaching consequences for the sustainability of many natural ecosystems in the coming decades. At the same time, we are plundering the aquifers of ancient water and extracting water from our major rivers to such an extent that sometimes they no longer make it to the ocean. In other words, we are disrupting those major cycles of the planet—the water and carbon cycles—and taking the Earth into uncharted territory.

The third difference is the scale of the human enterprise in which this current period of climate change is operating. Our planet's population is increasing; our cities are growing rapidly, often along coastlines; our world is increasingly and intricately interdependent—relying on global telecommunications, efficient transportation systems, and resilient provision of food, energy, and water. All these systems are already vulnerable to adverse weather and climate conditions; the additional pressure of climate change creates a new set of circumstances and poses new challenges about how secure we will be in the future. More than ever before, the world's changing climate is having a considerable direct and indirect impact on our livelihoods, property, well-being, and prosperity.

So we have every cause for concern and little time to act. As long ago as 1990, the former prime minister Margaret Thatcher recognized the threat that climate change poses:

> We can now say that we have the Surveyor's Report and it shows that there are faults and that the repair work needs to start without delay. . . . We would be taking a great risk with future generations if, having received this early warning, we did nothing about it or just took the attitude: "well! It will see me out!" . . . The problems do not lie in the future—they are here and now—and it is our children and grandchildren, who are already growing up, who will be affected.

Nothing that has happened in the last twenty-five years, including the increasing weight of scientific evidence, has detracted from the urgency of her message, and yet we still gamble with the future.

What we know and what we don't know

Temperatures have risen by about 1.0°C (1.8°F) since pre-industrial times; Arctic summer sea ice extent has declined by around 40 percent since records began in 1979; sea levels have been rising by about 3/32 inch a year since the early 1990s; each of the last three decades has been successively warmer at the Earth's surface than any preceding decade since 1850. We are more confident than ever that humans have been the "dominant cause" of the rise in temperatures since the 1950s (IPCC Fifth Assessment Report 2013).

Although we often talk about uncertainty in future climate projections, there are some things we can be certain about. We know that Earth will continue to warm; we know that the adverse impacts of climate change are disproportionately larger as we go to higher temperatures and that the risk of irreversible and disastrous changes increases; we know that sea levels will continue to rise long after we have stabilized the Earth's surface temperature and that melting of ice caps and glaciers will continue.

We also know that there will definitely be some level of climate change, whatever happens with future carbon emissions, because of the existing accumulation of carbon within the atmosphere. This means that some level of adaptation will be necessary whatever we do. The scale of the potential investments, for example in flood and coastal defenses, the risks associated with failure, and the long lifetimes and lead-times of such infrastructure together mean that future investments are likely to be highly sensitive to how the climate changes over the next two to three decades. We need to plan now for how we can climate-proof our lives, towns, and cities, and help to protect the natural environment.

The Earth's climate is immensely complex; the more we observe it and simulate it, the more we learn about the myriad of interactions that makes the climate evolve as it does. We have learned that the oceans, with their huge capacity to absorb heat, are fundamental to how climate change will manifest itself globally and regionally in the coming decades; we have learned that the response of the terrestrial biosphere—plants and soils—to warming and to changes in rainfall patterns will likely amplify the greenhouse effect from human carbon emissions by being less efficient sinks of carbon; we have learned that, through simple physics that tells us that warmer air holds more water, climate change will lead to more extreme rainfall and flooding events; and so much more. It is also sobering to realize that, with all these scientific advances, there is virtually nothing in the climate system that seems likely to dampen the effects of our greenhouse gas emissions; instead, the more we know, the more we are faced with the uncomfortable reality that this could be even more challenging than we previously thought.

In terms of assessing the risks of irreversible or dangerous climate change, we have made remarkable progress in recent years in building a new generation of climate models that simulate many more components of the full Earth system. This research is vitally important because observations alone cannot give us the future evolution of the whole Earth system or even tell us how it has evolved as it has. This knowledge is critical for deciding the pace and depth action to mitigate climate change.

In 2015 the Earth's surface temperature passed the 1°C (1.8°F) threshold—halfway to the 2°C (3.6°F) limit set in Paris—and yet we have already used two thirds of the allowable budget of carbon that we can emit if we want to stay within that 2°C (3.6°F) limit. Since then, new scientific evidence has indicated that this is an optimistic budget; the effects of melting permafrost and limitations on the ability of the biosphere to take up some of our carbon emissions suggest that we have even less room to maneuver than we thought. So if we want to gamble with the future of the planet we would be wise to take a cautious approach, because the odds are overwhelmingly stacked in favor of, at least, uncomfortable, and at worst, dangerous, climate change. Lack of certainty should not be used as an excuse for inaction.

Our climate in 2050

The year 2050 is not that far away in terms of climate timescales and we can paint a picture of what the world's future climate might be like with some degree of certainty. I won't go into an exhaustive list of facts and figures, but will try to draw out some of the human consequences of what our world might be like if we don't manage to reduce our emissions substantially. For instance, we know that the worst impacts will be felt by the world's poorest, who are already under enormous stress and have very few resources at hand to help them survive.

So let us fast-forward to 2050. The Earth's surface temperature has passed 2°C (3.6°F) above what it was just a century earlier and in the same period the global sea level has risen by another foot. The Arctic is now ice-free in summer, and there have been substantial increases in its ocean temperatures. Marine mammal, fish, and bird populations are changing, and the indigenous population is increasingly compromised by lack of food security; loss of coastal sea ice, sea level rise, and increased weather intensity are forcing relocation of some communities. The opening up of the Arctic has made it a major shipping route for international trade, and exploitation of the Arctic's natural resources is growing rapidly. New invasive species, brought in by increased human activity, are changing the natural ecosystems.

In India, pre-monsoon heat is now crippling for much of the population, especially across the northern plains, and flooding during the monsoon season is increasingly serious as daily rainfall intensities rise. Those living in low-lying coastal areas are experiencing more and more frequent incursions of seawater during storm surges as sea levels rise. Fresh water supplies are contaminated, agricultural land is damaged, and water-borne diseases are increasingly common. Forced migration is increasingly an issue. On the positive side, though, air quality has improved substantially and fewer people are affected by respiratory ailments.

Across the Tropics, construction and maintenance of infrastructure in major towns and cities have become more difficult as daytime temperatures frequently exceed thresholds where it is safe to work outside. Electricity demand for air conditioning is putting greater and greater pressure on supplies.

Several small island states, such as Kiribati in the central Pacific Ocean, are no longer habitable because of sea level rise, with the population now stateless and with an uncertain future; in others, coral bleaching has led to the loss of sustainable fisheries on which the population depend for their food security. Tourism, which was a major part of their economies, has fallen away.

Southern Australia and the Mediterranean, including the Middle East, are now in the grip of prolonged droughts and periods of extreme summer heat. Wildfires are becoming increasingly dangerous, threatening homes and urban environments, and damaging natural ecosystems. Water security is becoming more and more of an issue as aquifers are depleted. In the US, the weather is increasingly volatile with more extremes of temperature and rainfall. Summer heat waves are becoming more prevalent —eight of the ten warmest years ever recorded in the US have occurred in the last twenty years.

This small glimpse into what the climate of 2050 might be like is a stark reminder of why climate change will be such a determinant of our social and economic future, and of our role as custodians of the rich diversity of Earth's natural ecosystems. In all of this it is likely that water will become the most precious commodity on the planet. Understanding how regional rainfall patterns will change, the impacts of those on water availability and water quality, and the legal arguments of who owns water when rivers and aquifers cross national boundaries will be defining issues in the coming decades.

So far, debates on climate change have been dominated largely by uncertainty in the projections and the economic implications of dealing with the problem. But increasingly, climate change will become a moral issue. It is clear that the worst impacts will be felt by the world's poorest and that climate change has the potential to derail their socioeconomic development. As the UN Deputy High Commissioner for Human Rights Flavia Pansieri noted in 2015: "Human-induced climate change is not only an assault on the ecosystem that we share. It also undercuts the rights to health, to food, to water and sanitation, to adequate housing, and—for the people of small island states and coastal communities—even the right to self-determination." Looking forward, the protection of basic human rights and the role of the developed world in supporting the developing world are likely to alter fundamentally the debate on how we deal with climate change.

Our choices

The year 2015 was a landmark one: Not only did the world endorse the Paris Agreement (with the notable exception of the US—while technically still a part of the agreement, Trump announced in June 2017 that the country was withdrawing), but it also signed up to the Sendai Framework, which committed nations to work together for the substantial reduction of disaster risk and losses in lives, livelihoods, and health, and agreed on the Sustainable Development Goals to end poverty and hunger, improve health and education, make cities more sustainable, and protect oceans and forests. Combating climate change will be fundamental to achieving each and every one of these aims.

We cannot deal with global warming without accepting that the way we live has to change. We will need to transform the way we generate, store, and use energy, and we will have to learn to manage more extremes of weather and climate. One of the keys to this will be preparedness; continuing improvements in weather and climate forecasting will help us to be more resilient—forewarned is forearmed. Increasingly, our actions and responses to the natural environment themselves influence the environment and the impacts we feel from weather and climate hazards. For this reason, we need to make significant advances in the end-to-end evaluation of environmental risk from the original hazard, such as a flood, to the minimization of the risk by specific actions, such as new drainage schemes or planting trees, and assessment of the cost-benefit ratios. This will require integrating weather and climate simulations with advanced modeling of the built environment and ecological systems, as well as a better understanding of human dynamics and new approaches to modeling financial and socioeconomic factors. While the focus on climate change mitigation has largely been on reducing emissions of carbon dioxide, there are other atmospheric pollutants, such as black carbon and sulfate aerosols, and a reduction of their emissions would have significant benefits through improved air quality and hence improved human, plant, and animal health. China, where dealing with poor air quality is now a major driver in moving to a low-carbon economy, is a good example. But there are also risks that cleaning up those atmospheric pollutants that currently act to cool the planet might lead to an acceleration in near-term warming in some parts of the world and interact adversely with regional climate variability. So finding the best pathways to reduce our emissions in order to mitigate climate change is complex and will require an integrated approach, from the latest Earth system science to socioeconomic assessments.

Another factor in the debate on mitigation is the potential role of geoengineering. This involves the deliberate large-scale intervention in the Earth's natural systems to counteract climate change, either through solar radiation management or the direct removal of carbon dioxide from the atmosphere. It is increasingly clear that to achieve the temperature goals set by the Paris Agreement may require the wholesale removal of carbon from the atmosphere through the use of bioenergy with carbon capture and storage (BECCS). These techniques would have to be implemented on a global scale to have a significant impact on carbon dioxide levels in the atmosphere.

For solar radiation management, which considers, for example, modifying clouds with fine particles to make them more reflective or the injection of aerosols into the stratosphere to reflect more solar radiation back to space, we will need to properly understand the potential impacts. We must look beyond the effects on global surface temperature to the implications for the regional climate and hence, for example, to water and food security. These are not well understood and will require the same level of scientific diligence as has been devoted to understanding the regional consequences of greenhouse gas emissions. These techniques are particularly concerning because of social, legal, and political issues; they could be implemented unilaterally by nations without due consideration for their global impacts, and mechanisms for international governance still need to be put in place.

In the end, whichever way we go in combining adaptation (dealing with the effects of climate change) and mitigation (trying to minimize the level of climate change) it will be up to engineers and technologists to come up with innovative solutions to the problems of clean energy generation, storage, and distribution within a few decades and with as little disruption to the global economy as possible. At the same time, as individuals, we will need to find ways to protect the natural environment and ensure that we fulfill our responsibilities as custodians of life on Earth for future generations. This means that those of us that can will need to make choices about how we live and where we live, while still supporting those for whom that choice has been denied.

Some final thoughts

There is no doubt that climate change will affect us all profoundly in the future, but it's worth remembering that we do not go forward blindly without any sense of what we may be facing. The construction of computer models that simulate the Earth's climate and enable us to predict, from fundamental physical principles, how the weather and climate will evolve is one of the great scientific achievements of the last fifty years. In few other areas of science are we able to look to the future with the level of confidence that we now have in our climate predictions.

It is worth reflecting on the words of Vice-Admiral Robert Fitzroy, captain of the *Beagle*, who took Charles Darwin on his momentous voyages, but who was also the founder of the UK Met Office and who issued the first public weather forecasts. After the loss of the *Royal Charter* in a terrible storm in 1859 he wrote to *The Times*: "Man cannot still the raging of the wind, but he can predict it. He cannot appease the storm, but he can escape its violence, and if all the appliances available for the salvation of life [from shipwreck] were but properly employed the effects of these awful visitations might be wonderfully mitigated."

Over 150 years ago, Fitzroy embarked on the long journey of making forecasts as a means of reducing and managing the impacts of severe weather, and these now also apply to how we will manage climate change. From the global to the local and from hours to decades, our understanding of weather and climate and the predictions we make enable us to plan for the future and serve to keep us safe.

Let's leave the final word, though, to the British-born astronaut and climate scientist Piers Sellers, who died of pancreatic cancer in December 2016. On receiving his diagnosis a year earlier he wrote a moving piece in the *New York Times* on his perspectives on climate change:

> New technologies have a way of bettering our lives in ways we cannot anticipate. There is no convincing demonstrated reason to believe that our evolving future will be worse than our present, assuming careful management of the challenges and risks. History is replete with examples of us humans getting out of tight spots. The winners tended to be realistic, pragmatic, and flexible; the losers were often in denial of the threat. . . .
>
> As an astronaut, I spacewalked 220 miles above the Earth. Floating alongside the International Space Station, I watched hurricanes cartwheel across oceans, the Amazon snake its way to the sea through a brilliant green carpet of forest, and gigantic nighttime thunderstorms flash and flare for hundreds of miles along the Equator. From this God's-eye-view, I saw how fragile and infinitely precious the Earth is. I'm hopeful for its future.

THE FUTURE OF US

Medicine, genetics, and transhumanism

4

The future of medicine

Adam Kucharski

On April 26, 2016, a microscopic new threat appeared in Pennsylvania, USA. While the nation's attention was on the presidential primary being held in the state that day, a woman had arrived at a health clinic with symptoms of a bacterial infection. Doctors took a urine sample, and tests showed *E. coli* to be the culprit. But when the isolate was sent for further testing, it turned out that this was no ordinary strain of *E. coli* bacteria.

Given concerns about drug resistance, the local laboratory had just started testing samples for resistance to an antibiotic called colistin. Discovered in 1949, colistin is a so-called last resort drug. It's not commonly prescribed anymore because of the damage it can do to people's kidneys and is therefore used only against bacteria that can't be treated with other, less harsh antibiotics. And that's what made the sample from Pennsylvania unusual: It contained a gene that made the bacteria resistant to colistin. Although health agencies had spotted the gene elsewhere in the world, it was the first time it had been seen in the US. Fortunately, the Pennsylvania sample wasn't resistant to all antibiotics, but it demonstrated that such infections are chipping away at our lines of defense. "It basically shows us that the end of the road isn't very far away for antibiotics," said Tom Frieden, director of the US Centers for Disease Control and Prevention at the time.

Fifty years earlier, such statements might have seemed absurd. The 1960s was a time of optimism. Antibiotics such as penicillin had been used widely and effectively for almost a decade. Albert Sabin had developed a polio vaccine that could be taken with a sugar cube. Diseases such as tuberculosis were finally becoming curable. In 1967, William Stewart, the US Surgeon General, even claimed that "the war against infectious diseases has been won."

Yet the war is still being fought today. Despite intense vaccination campaigns, polio has not yet been eradicated. Drug resistant "superbugs" are making penicillin ineffective and tuberculosis deadly again. Meanwhile, we face the ongoing specter of a new pandemic—perhaps flu, perhaps something else—and the wider health challenges of an ageing population. However, we are also seeing remarkable developments in medical science, from genetics and personalized treatment to regenerative medicine and long-distance surgery. So where does that leave us? Should we be optimistic or pessimistic? What will the future of medicine look like?

The next epidemic

In medicine, there is always potential for surprises. Sir William Osler, who pioneered modern medical training in the early twentieth century, once described his field as an "art of probability and a science of uncertainty." Take infectious diseases. It is a near certainty that we will see another major viral pandemic within our lifetime. As for where, when, and which virus, we can only take an educated guess.

Osler himself died in 1919 during the infamous "Spanish flu" pandemic, which killed more people than the whole of the First World War. So far in the twenty-first century, there have been several new viral threats, including SARS in 2003, influenza A/H1N1p (aka "swine flu") in 2009, and Ebola in 2014. Though serious, all three had biological features that worked in our favor. Patients with SARS or Ebola generally showed symptoms when they were infectious, which meant that health agencies could track down people with whom they'd had recent contact and quarantine them, bringing the outbreaks under control. Flu is much harder to track in this way, but luckily the 2009 strain was far less deadly than the devastating 1919 variant.

This luck was important, because during a new outbreak we often lack effective drugs and vaccines. The problem is one of timescales: Research traditionally takes years or decades, whereas epidemics can last mere months. This is why there is still no effective SARS vaccine, and the vaccine against swine flu arrived late in 2009, well after the peak of the outbreak had passed. But things are changing. During the 2014–15 Ebola epidemic, researchers fast-tracked vaccine development. In less than a year, a clinical trial led by the World Health Organization had identified a highly effective Ebola vaccine. The next step is to have the research ready to roll out when a new infection appears. That means having a stockpile of drugs and vaccines that have gone through initial safety tests, as well as research teams that can carry out clinical trials at short notice and in difficult outbreak conditions. In 2017, a collection of governments and biomedical foundations launched the Coalition for Epidemic Preparedness Innovations to help tackle the problem. It aims to put $1 billion of funding into vaccine development, and will initially focus on three infections: Middle East respiratory syndrome, Lassa fever, and Nipah virus. All of these viruses have spilled over into human populations from infected animals but have not caused a major epidemic—yet.

Even if we have vaccines ready to test against a new disease threat, we still have to spot an outbreak when it appears. When it comes to controlling epidemics, our future success—or failure—will depend a lot on how we gather and analyze data. As well as testing for infection, for example, we can now collect and compare the genome sequences of viruses or bacteria collected from patients. At the start of the Ebola epidemic in 2014, this process took weeks; by the end, field teams could sequence viruses in hours on a device the size of a USB stick. From Ebola to flu, it's likely that every infection will be routinely sequenced in the future to understand how different pathogens are spreading and evolving. This will be particularly important if we want to track mutations that make current drugs or vaccines redundant.

Drug-resistant infections are likely to dominate medical headlines in coming decades. With overuse of antibiotics in humans and farmed animals leading to bacterial infections that cannot be treated, familiar procedures such as caesarean sections and hip operations could one day become extremely risky. It's not simply a matter of discovering more drugs; the newest antibiotic currently on the market was discovered back in 1987. Antibiotics are expensive and difficult to develop, but most patients need to take them for only a week or two, which means pharmaceutical companies have increasingly focused their research on other drugs instead. Tackling resistance to antibiotics will therefore require better stewardship of existing treatments, which means changing populations' attitudes and behavior. But these things can be tricky to study. What shapes your views about health? How does your behavior affect your risk of illness? What would persuade you to change your approach to treatment? Addressing such questions will require a mixture of biomedical and social sciences.

A better understanding of behavior will also be important for tackling other types of infection. For instance, it's likely that a reduction in risky behavior helped bring Ebola under control in West Africa. Many infections originated with family interactions and funerals; when this behavior changed, disease transmission often declined. However, we still don't fully understand how and when these changes happened in West Africa, or how significant they might be in the future.

Health researchers are increasingly examining how people move around the world and interact. They do this through surveys, mobile phone data, and satellite imagery. Soon it will be possible to link this data with other information—from genome sequences to environmental analysis—to study infections across a range of scales. Rather than focus only on the biological features of a disease, or its impact on a particular population, we will be able to simultaneously analyze the infection, its evolution, and its environment, as well as the behavior of its human patients. This will allow health agencies to design disease control strategies specifically for different populations and areas, and it will be particularly important in situations where a person's history of infection can influence their future risk of disease. Dengue fever is a good example; if you've previously been exposed to one strain of dengue, it can make your second infection more severe. That's why a 2016 study coordinated by the World Health Organization recommended that dengue vaccination campaigns should account for the history of infection within a population. At present, this requires collecting large numbers of local blood samples and carrying out time-consuming laboratory testing. However, new blood testing techniques are already making it easier and cheaper to identify what pathogens have previously circulated in a population. As more data becomes available, these tailored approaches will eventually become standard for every disease, in every country.

Bespoke medicine

Customized methods will become common in other areas of medicine, too. In 2015, US President Barack Obama launched the Precision Medicine Initiative. The aim was to develop treatments that accounted for a patient's genetic profile, environment, and lifestyle, rather than taking a one-size-fits-all approach. It was part of a wider trend in medicine, with methods increasingly focused on the specific patient as well as their condition. Although procedures like blood transfusions already account for individual variation to some extent (by considering blood type), these specific methods will use genome sequencing and other new tests to make it easier to predict how certain treatments may affect certain patients. For example, some cancer drugs are effective only against tumors with specific genetic characteristics. Similarly, a cystic fibrosis drug called Ivacaftor works for around 5 percent of patients with a certain genetic mutation.

Precision approaches will make medicine less reactive and more proactive. Instead of treating illness once it appears, detailed data will help us tackle risks before they become a problem. Genetic tests already make it possible to predict hereditary conditions, but these typically focus on a single harmful mutation, such as the BRCA1 gene mutation. If a woman has BRCA1, it means she has around a 65 percent risk of developing breast cancer over the course of her lifetime. In this situation, it is possible to reduce the risk through preemptive surgery. It was the presence of a BRCA1 mutation that led to actress Angelina Jolie's widely publicized decision to have a double mastectomy in 2013.

Rather than just looking at specific genes, it will eventually become common to examine whole genomes. This will mean a lot of data: researchers at Cold Spring Harbor Laboratory on Long Island have estimated that by 2025 human genome data will require more computer storage space than YouTube or Twitter. However, the cost of sequencing a genome doesn't include the complexity of analyzing the data. In some cases, the link between a single gene and a particular condition—such as BRCA1 and breast cancer—has been well established in huge clinical studies. Ideally, we'd have a simple set of rules like this for everything: "mutation A in gene B causes disease C." Unfortunately, if multiple genes are linked to a condition, or if the condition is rare, it is much harder to assess the risks involved. And that can make it tough to decide on preventative treatment.

To illustrate the difficulty of interpreting medical test results—whether genetic or otherwise—suppose there is a condition that affects 500 out of every 1 million people. There is also a test for the condition, which is 99 percent accurate. If you take this test and it comes back positive, what are the chances you will develop the condition? Remarkably, the answer is a mere 5 percent. This is because for every million people tested, we'd expect 495 to test positive and develop the condition (not the full 500 that will be affected because the test is only 99 percent accurate). Meanwhile, 1 percent of the remaining 999,500 people, or 9,995 people, who won't get the condition will mistakenly be tested positive (again, due to the test being only 99 percent accurate). So, the total number of people who would test positive will be 495 + 9995. Of these, only the 495 (i.e., 5 percent) will actually have anything to worry about.

Remember, this is using an excellent test that is 99 percent accurate. If we have a less reliable test, things become even harder to interpret. That's why a diagnosis often includes other factors, too, such as family medical history. Currently, patients who face such choices about genetic conditions can get advice from genetic counselors, who explain these risks and what they mean. Counselors typically provide information rather than recommendations. However, as genetic testing becomes more common, options may become more complex and the choices more difficult, particularly if there is limited treatment available. Would you want to know if you were at risk of developing an incurable condition? How would you handle the uncertainty of inconclusive test results?

With patients gaining more control over their data, the responsibility for making decisions will gradually shift away from doctors.

In the future, it will be up to us to manage our predicted health risks. We will have to choose what data to access and how to act on it. This will mean balancing potential benefits against the harm that might come from a misdiagnosis. In 2016, doctors at the Mayo Clinic in Minnesota published a report of a man who'd previously had a heart defibrillator surgically implanted after a genetic test suggested that he and his family carried a dangerous mutation. When they'd come to the clinic for a second opinion, it turned out that the test results had been misinterpreted: In reality, he faced no significant risk.

Patients aren't the only ones who will be interested in the results of genetic tests. In the US, health insurers are currently not allowed to deny coverage based on a gene mutation, thanks to the 2008 Genetic Information Nondiscrimination Act. But things are different for life insurance companies: Some have refused applicants who have the BRCA1 mutation. Unless the law keeps up with the rapid advances in predictive medicine, genetic health risks could also start to influence access to things such as housing and employment.

As our knowledge of hereditary conditions develops, it may also blur the definition for a patient. In 2015, a woman took St. George's National Health Service Trust in London to the British High Court because they had not told her that her estranged father had Huntington's disease, a genetic condition. Doctors had advised the father to inform his daughter when he'd been diagnosed in 2009, and he'd refused. The daughter, who had been pregnant at the time, would only find out that she had the condition in 2013. Should doctors have extended individual patient information to a family member in this situation? On this occasion, the judge concluded that they were correct to maintain confidentiality. It was the first time a court had ruled on whether medics should disclose genetic risks to family members, but it is likely that such issues will come up more and more in the future.

Families will also play an important role as the focus of medicine moves from acute to chronic illnesses. Conditions such as heart disease, diabetes, and obesity have long been a problem for well-off countries; now they are on the rise in low- and middle-income regions, too. Between 1975 and 2014, obesity in men rose from around 3 percent to 11 percent globally, and from 6 percent to 15 percent in women. Ironically, improved medical treatment will also contribute to chronic illness, because many conditions that are currently fatal will gradually become manageable, if not preventable. In coming years, researchers will find new ways to slow or halt degenerative diseases such as Alzheimer's. Implantable devices will monitor patients and tweak treatments accordingly. Stem cells will help fix or replace damaged body tissue. Last year, a team at the RIKEN Center for Developmental Biology in Japan announced they had successfully grown mouse skin in a lab, complete with glands and hair follicles. Rather than using skin taken from another part of the body to treat injuries like severe burns, it may eventually become common to use newly grown skin. Meanwhile, other groups have been developing lab-grown tissue to repair everything from bladders and corneas to ovaries and blood vessels. Improvements in healthcare already mean we are living longer than ever before; as it becomes possible to survive with a wider range of medical conditions, society will need to adjust to cope with the cost—both financial and emotional—of providing this long-term care.

A more connected world

The structure of society will affect our health in other ways, too. In 1970, only two urban areas had more than ten million inhabitants: Tokyo and New York. Fast-forward to 2017 and there are thirty-seven such "megacities." Many people living in these cities do so in poverty, and this is likely to get worse in the coming years:

the UN has estimated that up to two billion people could be living in slums by 2030. Infections like Ebola and dengue fever can spread quickly in these impoverished and densely populated environments, which can sometimes catch health agencies off guard. For decades Ebola wasn't seen as a major health threat. It had caused two dozen small outbreaks of hemorrhagic fever in Central Africa, mostly in rural locations, but there was little suggestion that it could ever cause an epidemic. Then, in 2014, this same infection hit three cities in West Africa and behaved very differently. We may soon discover other such pathogens that struggle and stutter in rural locations, but spread quickly in built-up areas.

Large cities will bring other health risks as well. Research led by the US Institute for Health Metrics and Evaluation suggests that over 5.5 million people die early each year as a result of air pollution. More than half of these deaths are in China and India, home to twelve megacities between them. As well as getting larger, cities will also become more connected. Every day, over 100,000 flights depart from airports around the world. This is why the 2009 pandemic flu virus could circumnavigate the globe within weeks, and why that colistin-resistant strain of *E. coli*—first identified in China in late 2015—could crop up in Pennsylvania a few months later. The flight network is also changing the geometry of infection: Traditional world maps, which show distance as the crow flies, can give a misleading picture of how we are really connected, and how contagion might spread. Once flight paths are taken into account, some cities may be much closer (or farther) than they initially appear.

Medical knowledge will also spread around the world more easily in the future. During the Syrian Civil War in 2016, experienced surgeons in places like Britain and Canada used webcams to guide local medics performing operations in the warzone. With improvements in augmented reality (AR) and robotics, more complex surgery could soon be carried out remotely. Worldwide sharing of data will also help with treatment at home. Rather than trying to book face-to-face appointments, patients may eventually discuss problems with artificially intelligent triage software, and if needed have an immersive AR consultation without leaving their house. From MRI scans to microscope slides, it will also become possible to make diagnoses using pattern recognition software honed on a global patient database. Doctors could then judge the risks of a particular treatment by instantly comparing their patient's condition with vast numbers of similar cases. And when a new infection appears—be it Ebola or *E. coli*—health agencies could quickly see how it is linked to other nearby outbreaks, and hence work out how it could be controlled.

Which brings us back to the question of whether we should be optimistic or pessimistic about the future of medicine. Even if we improve defenses against current diseases—eradicating things such as malaria, polio, and measles—other ailments will likely take their place. Whether it is drug-resistant bacteria or chronic illnesses caused by our changing lifestyles, there may be a limit to how far we can bolster human health, even in the distant future. The connections that link our world will eventually bring new health threats, and these will generate a host of biological, technological, and social challenges. Yet the increasingly global nature of medicine will also generate new ideas and approaches, allowing us to keep pace or even get ahead of some of these challenges. Ultimately it is our interconnectedness, and the collaboration it brings, that will also be the thing that saves us.

5
Genomics and genetic engineering

Aarathi Prasad

> Biology will be the leading science for the next hundred years . . .
>
> - FREEMAN DYSON, 1996

Genetics is the study of DNA—of the short sections of DNA we call genes, and of all the things that control them. Genes are our units of inheritance, the body's program files. They contain instructions—code to be executed. Flanking the genes are other files of code that define when the gene should be operating and how hard it should work. For most genes, their code is read and translated into proteins, which are the building blocks of our tissues and organs, our hormones, antibodies, and transportation systems in our cells, among other things. In this way, genes dictate an important part of our physical form, our health, and our potential.

Genetic engineering is what we have called the manipulation of genes. This kind of manipulation has a long history. Low-tech fiddling with genes (a.k.a. selective breeding) has turned ears of maize that looked like grass flowers some ten thousand years ago to the succulent closed quadric shape of the corncob, and it has turned wolves into pugs. More recently, through laboratories, it has created plants that are factories for pharmaceutical drugs, and produced a rice strain that carries a source of vitamin A, critical for our immune system and our eyesight. We are still manipulating genes for therapy, for fighting infection, and for the creation of new sources of energy and sustainable sources of food, but the way we do this has entered a whole new world of advanced technology whose potential has not yet been fully explored.

Genomics is a way of collecting and mapping complete gene data sets. It is an information-gathering field that grew out of genetics and molecular biology—sciences concerned with proteins and genetic material: DNA and RNA. Genomics thus seeks to catalog the make-up and machinations of the molecules of life inside our cells. Just over a decade ago, the first read-out of the chemical bases that make up the full human sequence of DNA was delivered. Represented by different combinations of just four letters, A, C, T, and G, this is our genome: all of our genetic material spelled out in a string of three billion letters containing about twenty thousand genes. Written out as a continuous sequence of letters of the size of the fonts on this page it would span the distance between London and Rio de Janeiro. The human genome is now part of a library that includes just over four thousand organisms, from bacteria and viruses to chimpanzees, duck-billed platypus, and Japanese canopy plants (one of which, by mass, contains fifty times more DNA than ours). But accumulating catalogs of coded life is not the end game. Taking as an example the application of genomics in medicine, the hoped-for pathway looks something like this. First, we must understand the structure of the genome and collect other relevant catalogs to show everything that is going on to make genes work. Then, we must work on understanding the biology of the genome—what the instructions all mean, which parts of the genome are indeed instructions, and which are not—and if they are not instructions, to figure out why they were so important that they remained a part of the genome over evolutionary time. Thirdly, we must work out what this map tells us about the biology of disease to address a number of questions. How and why, for example, does our DNA avert, or perhaps even precipitate, certain illnesses? How does our DNA interact with that of the microbes that can attack us? Finally, how can we use this enhanced understanding to develop targeted medicines that can be made to work smarter, with the hope of improving the effectiveness of healthcare?

Only within the last two decades have we gained the basic ability to read our genome and begun to figure out what our genetic instruction manual says. We have yet to find all the functional parts of the human genome, whether they form genes or not. (Some sections of our DNA, for instance, are there only to ensure the genome is stable or correctly organized.) This is where a significant part of the future of genetic research lies. Still, there is much that can already be interpreted. Once the completed human genome sequence was published in 2004, it took only two years for the first whole-genome test to become available on the market. Between 2007 and 2014, half a million consumers in the US alone bought home testing kits to gain access to the information carried in their genomes. Globally, through online and off-the-shelf sales, that market is showing no signs of slowing. Over the next two decades, we will not only be reading our genomes with proficiency, but we will also be digging deeper; mining genetic information we already have on record; breaking the security of former anonymity more and more adeptly—for better or for worse.

And, as we go beyond simply fiddling with the genetic codes already present in life on earth, a new age of true engineering of genes is just beginning. Because once we can read the genetic code, and begin to master its grammar, we can start writing it for ourselves. In the future, we will attempt to design DNA that is more and more complex, opening a window that within the next two decades will undoubtedly see the creation of new life-forms on the planet, or the resurrection of those that have become extinct. And, though it may sound less momentous, both philosophically and evolutionarily, the use of biological code outside biology will become an everyday part of the work and home spaces we inhabit. After all, DNA is a code that replicates with autonomy, one that for nearly four billion years has been unrelenting in its response to a changing environment, adapting and diversifying without human instruction. By understanding the genome, and then attempting to harness what DNA does with ease—self-learning, reactive adaptation, self-replication—things that non-biological machines struggle with today, the interface of computing and the machinery of DNA will be revolutionized.

Genetics and genomics of the future hold the potential to empower us through a greater understanding as well as the capacity to manipulate our bodies and our environments, and through greater personalization of technology. So how will we use these advances, and how will we use our growing knowledge of genes and genomes? Here, then, are some of the technological revolutions that are shaping the new genetic age.

Detecting the genome

Not so long ago, even analyzing DNA—let alone sequencing whole genomes—meant using large, expensive machines, housed in dedicated laboratories. The draft sequence of the human genome famously took fifteen months and cost around $300 million in the year 2000. By 2006, a draft of an individual's genome cost $14 million. Ten years after that, it cost about $1,500 and could be completed within two days. Now, sequencing is beginning to break out of laboratories, and even out of the hands of geneticists. Rather than requiring massive machinery, the new sequencers can fit inside your pocket. These devices cannot yet map your entire genome, but watch this space. Right now, they are being developed to study those genes known to be important for particular reasons, and doing so in real time. Recent innovations can detect genomic data, for example, of a bacterium, or strain of virus infecting a person. This is already showing promise for public health challenges in the places they are needed the most. In West Africa, one such portable sequencing machine was used to successfully identify 148 Ebola virus genomes from patients. In the field, in future, this type of miniature, user-friendly genomic technology aims to diagnose within hours, and guide the treatment of viral infections such as corona, dengue, Ebola, chikungunya, and Zika. Technology like this may one day be in everyone's pocket much like a mobile phone.

On an even smaller scale, we are starting to use semiconductors to detect DNA. Chris Toumazou, of Imperial College London, recently developed a chip that can be inserted into a USB stick that, within minutes, provides results viewable on any computer. Intentionally steering away from the full three billion chemical bases of the human genome to the 1 percent that each of us differ by, his semiconductor tech is creating our own "biological IP addresses." Different chips in USB sticks look at specific human gene mutations—to check our predisposition to a disease, say, or how well an individual is able to metabolize certain medications. As Toumazou puts it, "It will no longer be about a doctor looking at your medical history, but a doctor looking at your medical future."

Using the genome

While the potential opened up by the genomic revolution could be used to improve human health, it will also go far beyond this. The instructions for every life-form are encoded in its genome. In the last few years, new, smarter techniques for manipulating genes have come to the fore, and have generated a range of future prospects that could well have remained in the realm of science fiction. Much of this has been the result of a remarkable innovation by Jennifer Doudna and Emmanuelle Charpentier, who recently invented what has become the world's most powerful genome engineering technique. It is called CRISPR and was developed as a new system to manipulate DNA. It has allowed us to look more closely at the function of particular genes, by kicking them into action or shutting them down, or by cutting the DNA at precise locations to allow the sequence in the gap to be changed or added to. In theory, this means you could precisely remove a faulty version of a gene responsible for diseased cells and replace it with the "healthy," correctly functioning version. But this technology is also enabling new tricks beyond genome editing. In the future, it has the capability to help us figure out what the 98 percent of DNA that does not code for proteins is up to, which will take us beyond the alphabet of the genome to its grammar and punctuation. Part of this will be through our ability to turn genes on and off at will. Work has already begun on using light to control the location, timing, and reversibility of changes that we can introduce to the genome. This is an intersection with another fast-developing field, known as optogenetics—in which cells are controlled using light and which is currently being used in laboratories to deepen our understanding of the brain. In the next two decades it is hoped that the field will give us intriguing insights into a range of conditions, including neurodegenerative diseases such as Parkinson's, epilepsy, Alzheimer's, stroke, and hearing loss, which lack effective therapies at the moment. In addition, "CRISPR pills"—edible DNA sequences customized to instruct drug-resistant bacteria to self-destruct—are currently being developed.

We are already starting to see many other applications of CRISPR-related technologies—from the vital (preventing the loss of bee colonies) to the frivolous (dogs, mini-pigs, and koi carp as pets available in customizable colors). In the next few decades this technology could also create sophisticated biological circuits that convert cells into biofuel factories, for example by creating farm animals resistant to infection. In addition, there are plans for "CRISPi chickens," in which the tools for CRISPR editing will be integrated directly into the birds' genomes. Applications will include so-called "farmaceuticals"—like transgenic chickens whose eggs contain a drug to help with cholesterol problems.

There are also plans that could see the semi-"resurrection" of extinct animals. Work has started that looks at editing elephant embryos to make woolly mammoth–like arctic elephants and pigeon embryos to bring back a version of the passenger pigeon we hunted to oblivion in the nineteenth century.

Genetics, computing, and biohacking

Now that we can read it, it's easy to see how DNA, as a coding system, has parallels with computer science. The digital revolution of the last few decades has been a major facilitator of the genomic revolution and, in turn, DNA continues to connect with computing in a number of novel ways. What DNA provides, in no uncertain terms, is a form of data storage more long-lived than any of the storage media we have seen come and go over the decades. And unlike floppy disks or CD-ROMs, as long as there are people on the planet with the technology to read it, it will remain usable. An additional benefit of DNA data storage is size, since our entire genome fits into the nucleus of a cell, which is typically between two and ten microns across. To give a sense of this, a micron is one millionth of a meter, and even a human hair is about seventy-five microns across. As proof of concept, DNA storage has recently been used to record text files and audio files, including all 154 of Shakespeare's sonnets, and a twenty-six-second audio clip of the "I Have a Dream" speech by Martin Luther King. This was done by using a cipher-like binary code used in electronics and computing, but which uses the DNA bases A, C, G, and T in different combinations to represent bits (zeros and ones). A bit is the smallest unit of information that can be stored or manipulated on a computer, so for the Shakespeare sonnets 5.2 million bits of information were encoded into DNA.

Apart from data storage, DNA computing itself looks set to have a fascinating, if somewhat unintended, future. It started with solving a set of hard problems that computer scientists found interesting—because of the cost and technological limits of making increasingly miniaturized components for the ever tinier electronic computers, smartphones, and tablets that we've become reliant on. The ever-increasing miniaturization of our devices is about to reach its limit, however—Intel scientists predict this will happen as soon as 2018. But with DNA computers, individual molecules are the input, while other biological molecules, such as proteins, can act as processers. Unlike conventional computing, in which data is processed in sequence, which is time-consuming, DNA computers can process information in parallel, which dramatically speeds things up. Computations have been carried out using free-floating DNA or RNA in test tubes, or on glass plates, like a glass microscope slide overlaid with gold, arranged with a circuit-board appearance. A set of DNA molecules that encode possible solutions to a computational problem is created and attached to this surface.

They could also be used in places, and at scales, where traditional computers cannot go: inside a cell, or within very thin materials that we synthesize. This opens up new possibilities in which tiny bio-computers may be delivered into cells, to identify diseased tissue, selectively execute a self-destruct sequence, or else reprogram a cell's damaged DNA. For example, cancer cells might be reprogrammed so as not to multiply into tumors, or stem cells to grow into replacement organs. These bio-computers have also already been shown to be capable of controlled administration of biologically active therapeutic molecules, including some drugs.

Similarly, there have been some impressive examples of DNA-based molecular circuits—including those created to encode a version of the game tic-tac-toe (which interactively competes against a human opponent); another that calculates square roots; and neural networks that exhibit autonomous brain-like behaviors. So far, however, mimicking the digital logic of computers has been frustratingly elusive, and DNA computing power cannot rival existing silicon computers in executing any algorithm.

Still, the field is opening up exciting biological and biomedical applications. In the decades to come, DNA computing will create new and better DNA devices that can act as biosensors, be used in diagnosis, for nanofabrication, or to gain programmable control of our own cells' biology. It will also figure in the development and delivery of "smart drugs" to the places they are needed in the body: substances that will sense and analyze a variety of physiological cues and perform logical operations to release drugs or regulate how genes are expressed. And, apart from smarter living and well-being, work at the intersection of DNA and computing could also bring us closer to discovering the sequence of steps that happened in nature and through which life as we know it came to exist.

In the meantime, just as the large cabinets of mainframe computers that were once used only by specialists evolved into portable devices available to everyone, the prediction is that genetics will be increasingly accessed by people outside traditional laboratories and in more personal spaces. Who uses genetic science and genomics, and how and when, has already undergone a major shift, and this is set to become even more "democratized."

This brings with it both opportunities and challenges. Both doctors and patients are rapidly embracing the personalization that the genomic revolution is giving us, and personalized medicine is already being used in a variety of areas, from adjusting drug dosages to guiding the best treatment course for leukemia, HIV, and colorectal cancers. It has also become far more commonly used by people who are not ill, but who are nevertheless interested in exploring their future health. Many people are now using kits and services that can be bought off the shelf or online to access their genome through direct-to-consumer genetic ancestry testing. Andrew Hessel, a researcher at Autodesk whose projects include engineered nanoparticles, new DNA synthesis technologies, and mapping the human genome, says that the business model for DNA sequencing is shifting as we head into the future. There are already companies that will sequence a person for free, selling analyses or other services later. Others will pay to sequence the right people, because they have some valuable feature that would make it worthwhile for data miners to sift through their DNA code. What if genetic underpinnings can be found for desirable traits like not losing hair, not having it in the wrong places, or not greying; or for not needing much sleep, looking naturally youthful in later years, or having better eyesight or superior night vision? Investigating the genomes of people with these qualities may be a step toward finding a drug that could convey these traits to others.

But in order to glean any of our genetic information, samples of our DNA must of course be handed over—the secrets of our genome spat into a tube and posted to laboratories where small parts of it are read and interpretations sent back over the internet. It may sound harmless—even if you take your Viking ancestry or long-lost relationship to Cleopatra with the required pinch of salt, the limited interpretations of genetic data we are able to understand today are projected to increase to a level that presents a potential risk within the next decade. The human genome not only uniquely identifies its owner; it also contains information about ethnic heritage and predisposition to diseases and conditions, including mental disorders, that has serious implications for our privacy.

As computer security and privacy expert Emiliano de Cristofaro, of University College London, says, even recent public health initiatives aiming to create public datasets for "the greater good" require patients and donors to essentially forgo their privacy. Genomic anonymization—removing identifiers to conceal the identity of the genome owner—is hardly effective, he says. A team of researchers, for example, was recently able to discover the identities of people using only information available from genealogy websites. These are not even sites that have full genome sequence data, but merely small, disconnected fragments of these people's genomes.

De Cristofaro cautions that it's not just the development of computing power in the years to come that will make putting our DNA data out there even less secure, but also the fact that there are portions of the genome whose function we don't currently know and which may not seem risky to disclose. But, as we gain a better understanding of the genome, these portions of DNA data may well be used against their owners. It is not far-fetched to think that we will see the weaponization of DNA data. Knowing the peculiarities of a target's genome, it has been suggested, could be a great basis for the creation of personalized bioweapons that could take a target down and leave no trace. But on a more mundane level, as de Cristofaro asks:

> What happens if we discover that a mutation is linked to a mental disorder? Once the data is out, you can't take it back. Let's not forget that genomic data raises unprecedented security challenges, since the genome's sensitivity does not degrade over time and having information

> about a person's genome also implies information on their close relatives. And, after all, if your computer password is leaked it can be reset, but it's impossible to reset your genome.

In a less dystopian future, democratization will follow the path set in the last decade, in which community labs have cropped up in many major cities, where people with no scientific training push low-cost, accessible technology forward. People—non-scientists—have now worked on all manner of projects from producing cheap insulin to tinkering with food plants. But progress in genetic science is also presenting a space for leisure—anything from barcoding the DNA sequence of jams to making bacteria glow in the dark like a night light.

More than this, believes Andrew Hessel, our future is one in which biotechnology will come into almost every home. Just as we had personal computers, the Web, and now the Cloud, the future will see easy-to-use biotech devices in the home, such as toothbrushes that monitor mouth microbes.

While some of what the future of genetics will bring is already starting to come to fruition, other aspects may not be with us for some time, and may not happen in the way most people expect. But, as Hessel puts it, what we can be sure of is that DNA technologies are going to keep getting better, faster, and cheaper—and they are hugely important because they will allow us to design living things. "The use and rapid advancement of this technology is inevitable," he says. "We couldn't stop it even if we wanted to. Just like computers."

6

Synthetic biology

Adam Rutherford

We can probably agree that $314 is a lot of money for a necktie. Then again, this particular one has been woven from spider silk, which sounds legitimately weird enough, and might go some way to justify the cost. Spider silk is a remarkable material: a protein coming in a variety of forms, depending on species and the situation—for web-slinging, cocooning prey, or protecting their eggs. Each form has mechanical, elastic, and physical properties that exceed what humans can currently manufacture. Weight for weight, dragline silk is stronger than steel. Each form is squeezed out of the spider's abdomen via one of several spinnerets—a highly complicated set of internal spigots that align short fibers of silk protein depending on what type of web is required.

Much as we would love to harvest spider silk and exploit its remarkable physical properties, spiders are notoriously difficult to farm. Most have a range of personality traits that are not wholly conducive to the industrialized agriculture required for generating fibers in large enough quantities; spiders are largely solitary, and in the company of others tend toward cannibalism. So, the second reason why that tie costs so much is that it is a tie woven from actual spider silk that has never been anywhere near a spider or a spinneret. This silk has been grown and fermented in yeast.

Welcome to the strange world of synthetic biology. That title is of course an oxymoron: Biology is nature, and therefore cannot be synthetic. This conundrum is at the heart of a technology that will come to profoundly influence—maybe dominate—not just our sartorial future, but medicine, agriculture, drug design, energy production, and even space exploration. The necktie is the first merchandise produced by a synthetic biology company called Bolt Threads. In some ways it's a milestone, not really a commercially mass-marketable product—only fifty spider-silk-protein ties have been made, and though a stylish blue weave, they are pretty expensive. Bolt Threads is a company that is interested in manufacturing materials for clothing that bypasses the traditional forms of farming and harvesting materials such as cotton, wool, or silk from inefficient organisms such as plants and animals. We've been "designing" animals and plants for more than ten thousand years so that they produce goods for our own use. But farming has always been restricted to the slow, cumbersome process of breeding via sexual reproduction, and mostly limited by the biological necessity of breeding, almost exclusively via organisms within the same species. With synthetic biology, we can now bypass sex and breeding altogether and put together organisms not even slightly capable of having sex, having been separated on the evolutionary tree by hundreds of millions of years—spiders and yeast. Synthetic biologists seek to extract the source code and re-engineer it into much more efficient biological factories.

Nature's source code is DNA. The spider-silk protein is encoded in genes that for a few hundred million years have resided exclusively in spiders. A few decades ago, we began extracting and characterizing genes with accuracy and alacrity and started to wonder how we could insert them back into organisms, sometimes into the same species, sometimes into wildly different ones. The genes might be left intact to see how they would behave, or modified, or even deliberately crippled as a means of testing their natural function by seeing what would happen when some part was broken. This is genetic engineering, and it's enabled by the fact that the source code for biology is universal—all organisms on Earth are on the same family tree, an incomparably sprawling net that, according to Charles Darwin's idea of natural selection, has grown and blossomed for around four billion years. What that means is that all creatures are encoded by DNA. It's a deceptively simple alphabet consisting of just four chemical letters, strung together into a language that writes genes; the language of genes translates into words made up of amino acids. All living things use just twenty-one amino acids that link together to form proteins. Every protein that has ever existed has been written in this way, and all life is made of, or by, proteins.

The birth of genetic engineering in the mid-1970s recognized the fact that if done right, the cell or host organism didn't care where the DNA came from as long as it could read it. Genes that provided the instructions to make proteins and tell them what to do, which had been exclusively made by one species, could be written into the DNA of an entirely different species.

A few years earlier, The Beatles had invented sampling in music. On recording the track "Being for the Benefit of Mr. Kite" from *Sgt. Pepper's Lonely Hearts Club Band* (1967), George Martin and Paul McCartney had taken tape recordings of Victorian circus instruments called calliopes and cut them into short strips. The legend is that they threw them into the air, picked random ones off the floor, and inserted them into the recording. The notes were written in the same language, and by carefully aligning the key and the rhythm, the musical bridge section featured—for the first time in history—music physically lifted from another recording, seamlessly inserted to complete the track. Since then, sampling has become pretty universal in popular music, and has spawned entire genres, notably hip-hop, which has been the most lucrative musical form in history.

In 1973, a Stanford scientist named Paul Berg led a team effort that resulted in a gene being switched from one virus to another.

This act of biological sampling can be regarded as marking the birth of modern biology and the foundations of a biotechnology that would come to dominate all others. There are now no areas of the study of life that don't rely on these techniques. The identification of faulty genes that cause diseases, and then later the identification of all of our genes in the Human Genome Project (HGP), was entirely reliant on our ability to extract them and put them into bacteria where they could be manipulated, characterized, and understood. My own specific field of research—developmental genetics—relied on taking genes from one organism and inserting them into another that was better understood or more easily pliable to see what they would do. We took human genes and put them in bacteria, fiddled around with the code, and then spliced them into mice. In a couple of decades, from the 1980s onward, the act of remixing biology became normalized.

As the twentieth century gave way to the twenty-first, the fields of genetics, genetic engineering, and molecular biology grew from infancy into childhood. Those days, with the HGP bubbling away in the background, felt feverish and exciting, as I suppose all research areas feel when you're in the midst of a revolution. But in retrospect, we were all slow and inefficient, largely because so many of the manipulations of DNA that were the foundations of our experiments had to be invented each time we did them. We were cutting up tape recordings of calliopes every time we wanted to sample. This was a world of exclusively bespoke remixing.

This is largely the story of all new technologies. They have to be invented, which is messy and experimental. Then they are refined and widely copied and distributed so that everyone can use them. They quickly get easier and become normal. Sampling in music is now child's play, and is even possible on your smartphone. I type these words on technology unimaginable fifty years ago, a technology that I barely understand, that behaves largely predictably and that is based on simple principles of electronics and properties of materials. My action of hitting a key plumbs electrons through wires and circuits and logic gates and transistors and into light emitting diodes with such sophistication that I can't possibly fathom how it truly works. It was the commodification of parts that facilitated the boom in electronics. The components became standardized. You didn't have to invent a diode every time you wanted to use a diode. You could just buy one and stick it together with other components knowing full well how it would behave.

Component parts became smaller and design of ever more complex circuitry became easier. Nowadays, transistor-based electronics are involved in almost all aspects of almost all people's lives.

The founders of synthetic biology recognized this. Based primarily in the US, at Stanford and MIT, electrical engineers and mathematicians saw that genetics was coded circuitry and could be written and rewritten, but that geneticists spent half their time reinventing their circuits. If the parts of genetic engineering could be standardized like electronic components, then speed of progress in using biology as factories could increase exponentially.

So that's what they did. The BioBricks Foundation was formed in 2006 to create an open repository of standard parts of DNA, subtly modified so that they could be organized and bolted together like Lego. This was not a metaphorical assembly: The genes and genetic switches were extracted as strings of DNA, and each end molded so that it would stick to another piece in the correct biological orientation. Each of these genetic parts can be summoned from the repository, and can be sent around the world on a small piece of blotting paper. When you add a solution, the DNA simply floats off, and will join to the next component, like a domino. In such a simple act, you will have performed an act of genetic engineering unimaginable for almost all of history, and even in the history of modern science.

Frequently so good at predicting future technologies, none of the standard-bearers of science fiction saw this coming. Evolution, with its inimitable four-billion-year system of error and trial, has provided an unimaginable resource for building life-forms: genes in all of their innumerable forms, each one honed to optimize its own survival inside a host organism as the environment changes around it. With synthetic biology, we have invented a system that takes those component parts of evolution and remixes them not for survival of the host organism but for our own purposes.

There are many forms of synthetic biology, with some researchers not simply rewriting the code for specific purpose but reinventing the language of DNA with letters that don't exist in nature, or with modified versions of DNA that have never existed. Others have focused on utilizing DNA as a data storage device—after all, that is what it does for life anyway. Genes are information, and DNA is remarkably stable as a data format—we can recover the genomes of people and organisms that have been dead for tens or even hundreds of thousands of years. And because it's DNA, there will never be a time that we won't study it. Back in the digital hardware world, data degrades and formats go out of date in years rather than decades. Do you remember five-inch floppy discs? Or Betamax videotapes? Several projects around the world have successfully encoded videos, Shakespeare's sonnets, books, and other digital data into DNA, which stands currently as the densest form of digital storage known, at orders of magnitude higher than a Blu-Ray. At the moment, DNA as digital memory is slow to write, and slow to read, meaning that it's only of use in long-term archiving. But in the future, maybe we'll be running computers off DNA-formatted drives.

In this, the first decade after the birth of synthetic biology, the ideas have been extraordinary and the potential for inventing the future unprecedented. The range of ideas and techniques is astonishing, but the main drive has been to design genetic circuits that generate a product, often something hard or hitherto impossible for us to produce in any other way.

One of the higher points in the synthetic biology movement so far has been for the treatment of the single greatest cause of death in history. There are between two hundred and five hundred million malaria infections every year, and around four hundred thousand deaths, mostly of children under the age of fifteen. Throughout human history, this permanent specter has taken more lives than any other single cause. Treatments over the years have offered temporary respite, but unregulated overuse has iteratively resulted in the *Plasmodium* protozoa that causes the symptoms to evolve resistance and immunity to the specific drugs on offer. In recent years, a chemical called artemisinin has emerged as the most effective treatment. It has traditionally been extracted from the Asian sweet wormwood plant, but, as with much traditional farming, this makes it subject to economic boom-and-bust cycles, and the market price has fluctuated wildly over several years. In the first few years of the twenty-first century, a San Francisco synthetic biology company called Amyris was working on developing a form of diesel that could be grown in yeast, and one of the researchers noticed that one of the products in the tortuous molecular pathway was a precursor of artemisinin. So they began refining the genetic circuitry so that in yeast cells, artemisinin could be produced in large volumes and extracted as a novel source of the drug. The Bill and Melinda Gates Foundation recognized this as a potential strategy for wide-scale treatment of malaria outside the economic shackles of traditional farming and plugged in many millions of dollars, while a license was granted to the pharmaceutical firm Sanofi to bring it to market, at a significantly lower price.

Using cells as sensors has been under development since the early days of synthetic biology. The machinery of life can be incredibly sensitive—after all, the photoreceptors in your retina are capable of detecting a single photon of light. Today, cells have been reprogrammed to detect a panoply of environmental cues, from packaged meat going off in a supermarket to the detection of petroleum pollutants or infectious agents in the body.

And synthetic biology is not limited to Earth. NASA has expressed particular interest and has invested in developing technologies derived from reprogrammed DNA, primarily because cells are small and weigh next to nothing. The single biggest cost in space exploration relates to weight. Some estimates have suggested that it costs around $10,000 to break the shackles of gravity to get just one pound from Earth to space. If we are serious about putting people on other planets then we have to consider two key obstacles. The first is that space is a harsh environment—we have not evolved to survive outside the protective cocoon of the Earth's atmosphere—where spaceships will be bathed in enough cosmic rays and solar flares to induce a lifetime's worth of radiation exposure; on a simulated round trip to Mars, astronauts would suffer sterility and cataracts, and be at risk of developing a suite of cancerous tumors. The best way to counter those deadly rays is with thick metal shielding, which has a lot of mass, and therefore costs a prohibitive amount. NASA researchers based at Ames in California have been thinking about how synthetic biology could counteract these effects, and have been working on bacteria that secrete cytokines when triggered by radiation exposure. Cytokines are the body's natural response to DNA damage caused by radiation, so a synthetic bacterium that makes its own cytokines can counteract the damage.

The second reason NASA has a synthetic biology program is because once astronauts get to another planet they'll need shelter, oxygen, and food. Various projects have used bits of BioBricks to make circuits in cells that can generate oxygen, food, and even bricks—they secrete cementing molecules, and when grown in sand that emulates Martian regolith, they form bricks. The raw materials required for this are a test tube of cells, some water, and some Martian sand. Only one of those components needs to be transported from Earth.

This type of creativity has been entirely facilitated by an enthusiasm built upon the promise of synthetic biology's dreams. This is a field buzzing with youthful exuberance. Anyone can construct these genetic circuits—the only limits are those of your imagination. The problem has for a long time been that too few of these dreams actually worked. Circuits designed on paper didn't always behave as predicted once inside a living cell. What should by design have been clear digital outputs were often beset by noise in the system. Outputs, whether chemical sensors, drugs, or fuels, have all faltered. NASA's astronaut protection devices are decades from being usable, while Amyris's attempts to make clean biodiesel proved unscalable, and they couldn't get near to commercially viable volumes; artemisinin has been about to hit the markets on a large scale for a few years now, and has never quite made it. Black market artemisinin is already out there, and pockets of resistance to it are already emerging as users take it without following the World Health Organization guidelines, which instruct that it should be used only as part of a combined therapy.

Nevertheless, the promise of synthetic biology is very real. When it emerged in the first decade of the twenty-first century there was a very palpable sense of excitement, along with a period of hype in reporting and talking about it. But like so many emerging technologies, it didn't quite live up to expectations. It's my belief that we have now come through that hype period into a more realistic and settled era of real development where we will soon see very real, working products of synthetic biology providing services for very real-world problems. DNA has been redesigned as software—or, maybe more appropriately, wetware. As well as programming DNA, the wetware engineers are now looking at bug fixes for products that will soon hit the shelves. The Bolt Threads tie is a gimmick, a showpiece for a technology that will change profoundly how we make things. In science, revolutions tend to be slow and incremental. For tens of thousands of years we have grown, harvested, and refined the materials from which we built our world. Soon enough, much of those same things will simply be grown inside living cells with circuitry programmed by us—nature remixed.

7

Transhumanism

Mark Walker

Transhumanists believe that we should use advanced technologies, such as pharmacology, genetic engineering, cybernetics, and nanotechnology, to radically enhance human beings. In other words, we should be trying to create new types of humans—sometimes referred to as "post-humans"—who are significantly improved when compared with us. Imagine a future world populated by a new species of post-humans who are far happier, more virtuous, more intelligent, and whose lives are measured in centuries rather than decades: This is the future transhumanists imagine and work toward.

Radical happiness enhancement

We know happiness is important. We understand perfectly well why people forgo fame, power, or money if they think it will make them less happy, and why parents often say they just want their children to be happy.

Since transhumanism is devoted to making better lives, it is no surprise that transhumanists advocate using science and technology to radically enhance happiness.

We now know that the genetic component to happiness (defined as positive emotions and a general satisfaction with how one's life is going) is almost as important as the genetic component that governs our height. Of course, we need to interpret the idea of a genetic component to happiness carefully: genes are not destiny, and most human characteristics are formed by a combination of genetic and environmental factors. However, we all know people who are happier than most. These are the people who seem to have an extra spring in their step, who bounce back quicker from major disappointments, and so on. These "happy giants" are the owners of "happy genes"—gene sequences correlated with a happy disposition that have already been identified and located by scientists. As the biological sciences mature, it will be possible to create descendants who are far happier than the average human is today, using genetic engineering. Indeed, we already have the technology to at least begin making happier descendants. Creating "test tube babies" often involves genetic testing of the embryos created in a lab prior to being implanted into the mother. Technicians typically test for genetic markers for disease, and sometimes also for the sex of the child, but it is in principle possible that the embryos could also be tested—and selected—for genetic markers for predisposition to happiness.

But we need not wait for genetic engineering and genetic selection to make happy giants the norm. Radical happiness enhancement could come even to the genetically unmodified. My own "happy-people-pill" proposal seeks to put happiness-enhancing pharmacological agents into pill form to help those of us who have not won the genetic lottery for happiness (which is the vast majority). And unlike current pharmacological agents such as Valium and ecstasy, such pills would provide that naturally happy experience without any impairment of our cognitive faculties. If we were to direct societal efforts to the happy-people-pill proposal, we would have a good chance of creating such pills within a decade. Such a project would cost in the order of a billion dollars a year for ten years. A small sum compared with the size of the global economy, and the payoff would be huge. We could all walk with that extra spring in our step that the happiest among us now enjoy. And there are incredible secondary effects of making people happier. Happier people tend to be more social, to be better friends, to achieve higher academic grades, to have better marriages, to be rated more highly by coworkers and bosses, and so on.

Radical virtue enhancement

Enormous effort is put into making human beings more moral. Our childhoods abound with lessons about not hitting, not lying, playing nice and sharing, and we also educate adults about the requirements of morality, with, for example, public education programs about the evils of racism, and rock stars directing attention to the plight of the less fortunate.

A suggestion by some transhumanists, researched under the headings of "genetic virtue" and "moral enhancement," is that the efforts we put into the environmental side of making people more moral should be matched on the biological side. The proposal builds on two basic ideas: that human behavior and personality traits have a genetic component, and that morality involves behavioral and personality traits.

All behavioral or personality traits that have been studied appear to have some genetic component, and we can confidently predict that virtuous behavior also has a genetic component. In fact, it would be extremely surprising if it did not. We can even see the building blocks of virtuous behavior in other species; a mother mouse will care for her pups, and apes will sometimes share their food in a manner that we can only describe as just.

The genetic virtue proposal aims to provide our descendants with more of the biological building blocks that favor virtue, and fewer of the biological components that favor vice. As with happiness, these biological changes could conceivably result from genetic engineering, embryo selection, or advanced pharmacology. However, the same caution about genetic determinism applies here: If we create descendants with genes correlated with more virtuous behavior, there is no guarantee that they will in fact be more virtuous. At best, such procedures would increase the *probability* that they will act more virtuously. Genes are not destiny.

Radical life extension

The desire to radically extend our lifespan is probably as old as humanity itself, yet not much progress has been made toward this goal. At the time of writing, a French woman, Jeanne Calment, still holds the record for the longest human life at 122 years (1875–1997). Medical science hasn't yet been able to do much about the *maximum* lifespan, since we have good evidence that there have been a handful of people throughout history who have lived longer than a hundred years. However, it has done much to help the *average* age of humans, mostly by reducing infant mortality.

Radical life extension is the idea of using science and technology to push beyond this biological wall of just over a hundred years to allow humans, or post-humans, to live hundreds, if not thousands, of years. One way this might be achieved is to use the same strategy my neighbor uses to keep his old truck going: Simply replace worn parts. So, while the average car lasts about eight years or so, my neighbor has an immaculate truck five times that age. In theory, it will be possible to keep a human body going indefinitely so long as we repair and replace ageing cells and organs, perhaps by using advanced stem cell technology. Using a patient's own stem cells, scientists have already demonstrated that it is possible to grow new organs, like bladders, on a lab bench. The idea is to "cut and paste," for example removing an old bladder and replacing it with an organ grown in a lab. And it may not be necessary even to do this; in clinical trials, stem cells have been used to repair damage to heart muscles suffered during a heart attack.

Part of the difficulty in reversing ageing and death is that there are numerous causes involved. The SENS foundation, a nonprofit organization devoted to advancing radical life extension, has identified seven causes of ageing (cellular atrophy, cancerous cells, mitochondrial mutations, death-resistant cells, extracellular matrix stiffening, extracellular aggregates, and intercellular aggregates). Each cause will require its own biomedical or engineering solution. Still, the fact that we are making progress, as demonstrated by some remarkable techniques to extend the lives of animals, suggests that science and technology will allow us to radically extend our own lifespans. Recently, scientists were able to extend the lifespan of mice by 25 percent simply by developing a cellular "kill switch" that removes a certain class of old cells (senescent cells). Not only are the mice living longer, they are also healthier, with fewer age-related diseases.

None of the technologies described so far will help in the case of catastrophic injury; all the stem cell technology in the world won't help you when a piano falls three stories and lands on your head. But transhumanists have envisioned technological fixes for even these contingencies, which—though futuristic—might well be possible within the twenty-first century. The idea is, in effect, to create a backup: a possibility would be scanning and recording your brain down to the micro-level and then reconstructing a new brain from the data. One version of this proposal envisions a small army of nanobots—molecular size robots—recording the molecular structure of your brain. In the event of traumatic injury, the nanobots could be activated to repair a brain crushed by a piano or to create a new brain based on the saved blueprint.

Another possibility is to upload this information about your brain to a computer platform. In effect, this would be like moving your mind from biological "wetware" (your brain) to computer hardware. This raises a number of thorny metaphysical questions about whether it is really "you" who survives this migration, or whether this is simply an exercise in making a copy of you. Transhumanists (and philosophers) are deeply divided on this question. I side with those who say that we *can* survive such migration. Indeed, I go so far as to argue that there could be multiple versions of one's self; if you survive the transfer from a biological platform to a computer platform, then there is no principled reason why you could not be uploaded to multiple computers. In this way, immortality, or at least something very close to it, might be achieved.

Radical intelligence enhancement

There is a well-known correlation between brain size and intelligence. (Actually, for many species, the body weight of the species must be factored in as well, but we will ignore this here.) Our brains are almost three times the size of a chimp's, even though we are genetically very similar—around 96 to 98 percent of our genome overlaps. As the human genome is made up of about twenty thousand genes, this means that fewer than one thousand genes separate us from our cousins, and perhaps as few as four hundred genes. Using existing genetic technologies, we could try to create a post-human species. Let us call this species "*Homo bigheadus*," with a brain volume two or three times that of humans today. In terms of technological challenges, this is actually fairly straightforward. There are special genes (homeobox genes) that control the size of various body parts. By increasing the expression of genes that control brain size during the development of a human zygote, in principle it is possible to attempt to create beings with much larger brains. Using different combinations of these homeobox genes, we could increase the growth of certain other parts of the brain. For example, the neocortex is associated with higher cognitive functions, and so if the goal is to increase intelligence, it might make sense to try to increase this part of the brain by manipulating homeobox genes that control the development of this area.

But to say that we *could* attempt this is most definitely not to say that we *should*. We do not know if any such experiments would succeed in creating post-human intelligence—beings that compare with us in intelligence as we do with apes. But we also have no reason to think that such experiments will fail. This is why we do experiments in science: to find out. But today, when we use genetic technologies on plants, for instance, most experiments go badly. And it is one thing to throw a bunch of failed plant experiments into the compost bin, and quite another to do so with human zygotes or infants. Still, we already have the science and the technology: This is a line of experimentation through which we could start creating *Homo bigheadus* today. Thankfully, very few scientists are prepared to put safety and ethical and moral considerations aside in order to do so. Such controversial research will need to be discussed and considered carefully by society as a whole if it is going to avoid comparisons with discredited ideologies such as the eugenics movement.

There are other less dramatic means to attempt to achieve cognitive enhancement. Some studies have suggested that about one third of all college students use pharmacological enhancement—smart drugs such as modafinil or adderall—to increase academic performance. There have also been some initial experiments in the "cyborgization" of the human mind by creating brain-computer interfaces. Increased brain capacity might also be possible for human adults simply by creating a large number of new brain cells using neogenesis techniques. (Adult human brains tend to make only a relatively small number of new cells.) We know from experiments on Alzheimer's patients that neuronal fetal tissue can integrate with adult neural tissue and function successfully. So, this technique promises a potentially different way of creating *Homo bigheadus*. We could keep adding new brain cells (along with increasing the cranium size) of test subjects to see whether radical intelligence enhancement is possible.

Reasons for transhumanism

In thinking about reasons for transhumanism, we are engaging in the question of whether transhumanism is a morally desirable outcome, not about how likely it is that transhumanism will come about. This chapter considers a moral claim about the future: how the future *should* unfold, not how it *will* unfold.

While transhumanism claims that radical enhancement is a morally worthy goal, this does not imply that transhumanists believe that everyone—particularly those opposed to transhumanism—should be radically enhanced. Think instead of the claim that getting a degree is a worthy goal; it does not follow that everyone should be forced to go to college. Not only would this be a violation of individual liberty, but it is likely to be counterproductive. Transhumanists typically believe that it is wrong to force adults to do something, even if we think it is very good for them.

The flipside of this anti-paternalism is that transhumanists endorse the idea of a "right to biological freedom." Biological freedom is already recognized to some extent: we permit citizens to modify their biology with laser eye surgery and all sorts of cosmetic surgery, such as breast enlargement. The suggestion by transhumanists is that radical enhancement is, or should be, covered under this general right of biological freedom—to do with our bodies as we see fit. Indeed, we can see that biological freedom is intimately related to the freedom of conscience recognized by so many modern liberal states. Limiting radical enhancement of intelligence or happiness is to limit our conscience, for example, consigning us to fewer happy mental states, or more limited thinking.

Transhumanists often appeal to a "companions in innocence" line of argument based on an analogy between biological enhancement and cultural improvements; remember that it was only a few thousand years ago that efficient forms of writing were invented that permitted humans to store information in written work. Writing dramatically enhanced our intellectual capacity and our ability to recall information with the creation of additional storage space outside our brains. We tend to think that this cultural enhancement was a very good thing and, so the analogy goes, we should think the same thing about some biological enhancements.

Many transhumanist arguments for enhancement focus on the positives for specific types of enhancement. As noted above, the right or duty to enhance for happiness may be connected to the general desirability of happiness. The right, or even duty, to enhance for virtue is linked with the idea of making our lives and our world better. The aim of increasing intelligence is intimately connected to the goal of discovering the truth about the universe and our place in it. Finally, the goal of living longer is linked to the idea of having a rich and fulfilling life.

Arguments against transhumanism

Up to about the late 1990s, objections to transhumanism tended to focus almost exclusively on the unlikelihood of transhumanism rather than its desirability. The idea that technology could be used to radically enhance human beings was criticized (often simply lampooned) as being science fiction. After the birth of the first cloned mammal in 1996—Dolly the sheep—the tide started to change. Critics of transhumanism began to begrudgingly concede that some of the proposals of transhumanists might be technically possible, and turned to questions of desirability or ethical rightness.

Leon Kass, former chair of the President's Council on Bioethics, has suggested that transhumanism is a form of self-undermining hubris. Kass argues that effort and sacrifice are intimately intertwined with many of our deepest desires and goals. With transhumanism, obtaining these goals turns into questions about how technology can achieve these ends, rather than human effort. When Ferdinand Magellan (almost) circumnavigated the globe in 1521, it was an admirable triumph of the human spirit. Today, Magellan's route can be retraced by buying plane tickets. But such a plane journey could hardly be described as a triumph of the human spirit. However, the strength of Kass's objection will depend on how plausible we find the claim that we will not be able to find new goals to challenge ourselves should we improve human biology.

Francis Fukuyama worries that transhumanists would destroy the basis of equality of modern political life; we share a common human nature and this shared nature is the basis of political stability and legitimacy. Fukuyama famously rhetorically asks: "What will happen to political rights once we are able to, in effect, breed some people with saddles on their backs, and others with boots and spurs?" One answer to this question is that, just because we have the technology to create subservient people with "saddles on their backs" is no reason to think that we should or will create such beings. Indeed, the reason why we fear this outcome seems to be because we know that throughout human history, humans have exploited and enslaved other humans. But according to transhumanists, transhumanism with a genetic virtue proposal offers hope of escaping this tendency, not perpetuating it.

Elsewhere, New Zealand philosopher Nicholas Agar offers a philosophically sophisticated critique of transhumanism. Suppose advanced technologies were used to improve the cognitive capacity of my dog Coco. Coco's brain is enlarged to the size of a human brain, and she is provided with the neural architecture to support language functions. Her mouth, tongue and throat are modified to allow for the production of speech. Her front paws are modified: She is given opposable thumbs and digits to allow for fine motor control. As a result, she has the dexterity necessary for typing on a keyboard, threading a needle, etc. As Coco develops her newfound linguistic, cognitive, and behavioral capacities, she starts attending the local elementary school, and soon she is off to high school and college. Her old life seems pointless to her; she is no longer ecstatic at the thought of a walk around the block or sniffing lampposts. She is no longer content to sleep on a bed only when a human is not there, etc. As we think about the enhancement of Coco, it seems like a better description might be the destruction of Coco. The new Coco shares little of the concerns, beliefs, and desires of the old Coco; we changed her so much that she ceased to be. Agar's argument, in effect, is that transhumanism will not lead to the enhancement of human lives but to the disappearance of humans. This line of argument against transhumanism is reminiscent of Aristotle's suggestion that if one is wishing for what is best for one's friend, it would be wrong to wish that they become a god. Such a wish involves a change of such a radical nature that it would mean the end of the friend. In a similar way, Agar suggests that transhumanism does not answer human concerns and desires, but involves the creation of beings with non-human desires and concerns.

Many criticisms of transhumanism focus on specific proposals for radical enhancement. The idea of enhancing for happiness has been criticized as something akin to Huxley's *Brave New World*, where authentic happiness has been abolished. Moral enhancement is criticized on the basis that the enhanced would lose their autonomy or free will. And critics object to radical intelligence enhancement out of concerns that it might lead to the creation of "evil geniuses."

Transhumanists have countered with spirited defenses against all these charges. One line of defense is what we might think of as the "companions in innocence response." There are unaltered humans at present who are naturally happier, more moral and smarter than the average, thanks to the genetic lottery. Transhumanists argue that no one thinks such people lack authentic happiness, lack free will, or are likely to turn into evil geniuses. So, transhumanists argue, it is simply false to suppose that genetic enhancement will lead to such untoward consequences, either. To illustrate: Imagine meeting two individuals in the future who are happier than most due to a genetic predisposition. One individual's genome was genetically engineered, the other received her genetic predisposition through the genetic lottery. Unless you were told, you wouldn't know which was which, hence it seems morally arbitrary (to transhumanists) to say that one of these individuals lacks authentic happiness, for transhumanists reject the idea that there is an important moral distinction between the naturally and the artificially created.

Cultural considerations

Transhumanism raises important and profound questions about the future of humanity. Indeed, I hope the reader sees that transhumanism has much claim to being the biggest question of the twenty-first century. In a single chapter it is not possible to discuss, even in outline, all the areas potentially affected by transhumanism, but allow me to end by mentioning just a few additional considerations. One is the relationship between transhumanism and religion. While it may be thought that religions are necessarily antithetical to transhumanism, the Mormon Transhumanist Association, one of the longest-running and most active transhumanists groups, is one of several promoting the idea of deep compatibility between religion and transhumanism. Another issue passed over here is the question of the creation of artificial persons, like advanced computers or robots. And the question of the relationship between transhumanism and art has not been broached. The relationship between transhumanism and politics was mentioned only in passing in this chapter, but pro-transhumanism candidates have run for political office in several countries.

Finally, and perhaps most importantly, nothing has been said about the potential for global catastrophic risks. Transhumanists are as concerned about avoiding bad outcomes as they are about promoting good uses of advanced technologies, but it is true that many of the advanced technologies of interest to transhumanism could just as easily be used for evil as for good.

THE FUTURE ONLINE

AI, quantum computing, and the internet

8

The Cloud and "Internet of Things"

Naomi Climer

The Cloud and the so-called "Internet of Things" (or IoT for short) make a powerful combination, and one that will increasingly shape our lives in the future. Put simply, Cloud computing provides access to a shared pool of resources such as storage, processing, applications, and services. The idea is that you don't need to have all these capabilities on your own machine if you have reliable access to them elsewhere. Indeed, if you can access them whenever you need them, you'll be able to use much richer resources than you could afford to have personally. Once you put things in the Cloud, mobility follows. You can pick up your Gmail in Paris or Cairo, you can work in Chicago and Tokyo staying connected to the office HQ and watching Netflix in your downtime. This has already transformed flexible working, personal entertainment, and lifestyles.

The other thing that the Cloud makes easier is collaborative working. If data and applications are remote, it's feasible to bring people together to work on something at the same time without being physically together. This can be hugely important in industries where there are workers in the field who can now be virtually "joined" by experts supporting them without being on site. In the creative industries, broadcasters covering the Olympics used to send their whole team to the event, but now they can send far fewer people and do the bulk of the work back at base. This isn't just a case of getting everyone on the internet—it requires collaborative tools to help people work collectively in a virtual environment. (If you've ever tried to have a telephone conference with ten people, it can be very hard to manage who speaks when. Add some very simple visual online collaboration tools to this call, such as the ability to raise your hand, and the experience becomes much more meaningful.) The quality and creativity of virtual collaboration tools will be a key driver of future Cloud usage. But it's an indication of how, in the future, we will all be much more connected, and how central the Cloud will be to that.

If the internet is one network for the planet—something that allows people to collect and exchange data—then the IoT is something that allows *things* (physical objects such as cars, thermostats, fridges, etc.) to do the same. In 2016, InternetWorldStats.com estimated that there were about 3.7 billion people on the internet. There are already far more things on the internet than people, and estimates suggest that there will be about 50 billion things by 2021. Already, we're seeing smart (IoT) devices for fitness/well-being tracking, monitoring vulnerable people, and continuously monitoring for specific medical conditions such as heart rate or blood sugar. It's early days yet and there's quite a lot of debate about privacy, security, the social impact, and efficacy of these solutions, but they already seem to be an unstoppable force.

Meanwhile, and less visibly, industries are beginning to test the waters, and these may be where the greatest economic benefits will come in terms of contributing to big, global and national challenges. For example, wind farms find that they can use information from sensors to continuously adjust the blades to make the most of the wind. It's like tweaking the sails on a boat to get a little bit more speed—it can increase the amount of energy generated by up to 25 percent. Factories are beginning to connect their own data with their suppliers so that components are automatically ordered just in time for production. Sensors around the production line can also help to predict problems before they happen and delay scheduled maintenance when the machines are fine, making sure the engineering effort is being used where it is really needed. International manufacturing companies such as Bosch or Airbus are able to share data between factories to make sure that issues and fixes are managed globally using expertise from any location.

The other feature that the IoT allows for is the collection of data on extraordinary scales. Very cheap, low-power consuming, connected sensors can be added to everything to collect real-time data about individuals, the environment, or specific systems, allowing us to continuously check for leaks with sensors in remote water pipelines or to crowdsource vast amounts of data from mobile phones to track an outbreak of flu.

Actually, one of the challenges of the IoT is how to turn vast amounts of data from multiple sources into meaningful wisdom. The prevailing term around this is DIKW: Data—Information—Knowledge—Wisdom. The data might be a string of numbers. The information in that data might tell you that you have product number thirty-two with a sell-by date. The knowledge you can glean from the information is that product thirty-two is milk and the sell-by date is yesterday. The resulting wisdom is that it's time to buy more milk! It's a trivial example, but with vast and disparate data, it can be a real challenge to glean meaningful wisdom from the sea of information. Whole new job categories are opening up around data science as we seek people who have the talent to help work out how to turn unlimited data into useful wisdom. What are the right questions to ask of the data? What are the important patterns?

One future assumption is that the majority of the world's population will live in cities within a couple of decades, and it is clear that this dense population will bring social and practical challenges around energy, education, health, and transportation, among other areas. Experiments have already started with "smart cities"—urban developments that blend together multiple online systems such as hospitals, schools, and transportation—to overcome challenges and develop applications that will improve the quality of life. There are already many examples of smart city trials around the world. Trials have covered a wide range of applications, from beacons to help visually impaired people find their way around in San Francisco and Euston station in London to sensors in trash cans allowing for a more responsive, "smart" approach to garbage collection. There are now many "smart waste/recycling bin" solutions on the market, such as Bigbelly (US), Enevo (Finland), Ecube Labs (South Korea), along with active trash can checking in place in thousands of locations worldwide. The garbage systems will not only sense when a bin is full, they will also predict when it is likely to be full and suggest efficient driver routes. It seems that the technology works rather well, but it takes time for the drivers to trust the system and adjust their working practices.

This gets to the heart of a truth about the future of the IoT and "smart" technology: we will have to get used to having it in our lives. One of the fundamental strengths of the IoT is the way it can readily connect apparently unconnected things to create something new—something that could have profound effects on the efficiency and efficacy of services, from transportation to healthcare and education, if we manage to connect them. This is more of a practical challenge than a technical one—these industries and government departments are not particularly used to working together and it will take a sustained, collaborative effort to make it work. There are currently five governments (the D5) that are working together to strengthen the digital economy, bonded by the principle of openness. These are Estonia, Israel, New Zealand, South Korea, and the UK. Between them, they are championing open standards, international collaboration and open government, and collectively pushing the agenda forward.

Another way we're currently exploiting the Cloud and the IoT is virtual reality (VR) and augmented reality (AR). We've seen these in computer games and the commercial mainstream for some time, but now they are popping up elsewhere, too: Architects and estate agents are using VR to help visualize things that are yet to be built; companies building ships, airplanes, and oil rigs can use the technology to visualize things before making a massive investment; some factories are using AR as a way to overlay instructions on a headset for technicians in order to give them step-by-step guidelines on what they need to do. Both VR and AR are still relatively young, but their potential to become an everyday part of education, health, work, and leisure is massive.

Looking beyond what we already have, there are numerous possibilities for tackling some of the big global challenges with the Cloud and the IoT.

Healthcare advances will include more of the well-being sensors that we're already using. It's likely that we will have more sensors on our bodies monitoring our physical and emotional states and encouraging us to manage our own health more effectively, both day and night. Sports stars already make extensive use of body monitoring technologies to help them reach the optimal physical state to compete. Industries where employees have to cope with a lot of stress, such as financial services, are also experimenting with staff monitoring to help manage stressful environments. But we may one day see a world in which all of us are able to monitor ourselves constantly, receiving a live stream of data from our devices. Our bodily sensors might tell our fridge to order more fruit and vegetables, tell our central heating that we have a fever and need to be warm, and call the doctor when our temperature rises above a certain level. Our sensors may even reschedule tomorrow's appointments because they know we won't feel well enough to go to work.

The ability for remote diagnosis and care by healthcare providers will be improved, including supporting older people to remain in their homes for longer than may be possible today. Wearable technology such as exoskeletons and prosthetic limbs is likely to benefit from IoT connectivity because it will add new control mechanisms depending on physical capabilitics. (You could run your whole body from your tablet—the i-limb from Touch Bionics already does this for a prosthetic hand that can be controlled from a smartphone app.)

Analyzing the vast amounts of data generated by the IoT in order to find new patterns could be life-changing. For example, Stanford University ran big data algorithms on illnesses grouped by how they changed gene activity matched against medications blocking those pathways. They concluded that various existing drugs might prove effective against other illnesses in addition to those they were developed for. Given the expense and time involved in clinical trials of new drugs, this could be a big step forward, but it will require the data from clinical trials to be made available for everyone to use. Pharmaceutical associations in the US and Europe have now committed to sharing their data. In the future, we can look forward to similar breakthroughs.

Our healthcare is likely to become much more personalized, too—as the big data about us is harnessed for our benefit, a doctor's system is likely to analyze everything that is known about us from our lifestyle (social media, CV, sports played, subscription to wine club, etc.), to our genetics, our reactions to previous treatments, to all known worldwide research on the relevant illness. Of course there are significant privacy issues to be overcome here (of which, more later), but the potential for better outcomes in healthcare is compelling.

One of the benefits that the Cloud and the IoT are probably already able to deliver, but haven't just yet, is universal education across the planet. As our world paradigm changes, the opportunities to learn what you need to know and be ready to engage in the new economy no matter where you are in the world are excellent, in theory. The main constraint is getting basic internet to everyone—this is something that the United Nations and the World Economic Forum have talked about, and some of the tech giants such as Facebook, Microsoft, and Google, along with celebrities/philanthropists, are working on this.

The fully connected, virtual, efficient, data-driven, personalized world of the Cloud and the IoT has the potential to improve the quality of life for many people who currently struggle with disadvantage in many forms. It has the potential to reduce power consumption and food waste, improve production efficiency, and transform our healthcare experience into something much more tailored to our specific needs.

However, there are many challenges ahead of us before we can fully achieve this dream. The four that I'll particularly call out are security, coverage, energy, and society.

There is a wide range of issues surrounding privacy and security. For individuals, the concern is mainly data privacy. To get the most out of the IoT, you'll need to allow your devices to know everything it is possible to know about you. The more you keep from it, the more imperfect the service you will receive. To some extent, this will be a matter of personal choice, but significant safeguards will be needed as well as continued personal vigilance. Already, there is a lot of concern that the big tech companies such as Apple, Facebook, and Google, as well as governments, have access to extraordinary amounts of data about us. They use this data to personalize and improve their service to us, but it does raise concerns that the data can be abused. There's a fine line between personalizing services and spying! Your health data could be used to increase your insurance premiums; your lifestyle data could be used to check your tax return accuracy; your personal data could be used for identity theft if it is not properly protected. Big tech companies are becoming increasingly powerful as a result of the personal data they collect, and the direct access they have to billions of people and their vast incomes. Such power can be a force for good or has the potential to be used to exploit and manipulate. Governments and consumer groups around the world are wrestling with this issue today.

For both individuals and companies, there's the risk that any connected device is a gateway for a hacker. Could a hacker take control of your home by hacking into your kettle? Could a government take control of another's power plant by hacking into one of the connected machines? The answer to both of these questions is yes. One of the earliest major cyber hacking stories was Stuxnet in 2010, a worm virus written to very specifically target programmable logic controllers on the Iranian nuclear program's centrifuges. The virus caused the centrifuges to tear themselves apart and was reported to have ruined almost one fifth of Iran's nuclear centrifuges. In 2015, a cyberattack in Ukraine temporarily cut the power to eighty thousand customers. These two attacks suggest that the future may also see warfare being played out across the internet, and the IoT and Cloud may become weaponized by hostile governments.

Also in 2015, a group of hackers demonstrated taking control of a jeep on the freeway, having hacked in through the infotainment system. Since this "benevolent" hacking, the automotive industry has stepped up the efforts to protect connected vehicles from attack. The risk of a cyberattack to connected vehicles or homes, to national infrastructure such as power or water, or to key businesses, remains very real. Commentators say that this will be one of the ways that future wars are fought, so cybersecurity is likely to be a headline topic for decades to come.

Less dramatic is the question of coverage. A connection to the internet is an obvious requirement behind many of the applications of the IoT, whether broadband, wireless, or cellular. While this can often be managed in urban environments, it is far from guaranteed in remote areas. Autonomous vehicles, remote surgery, and remote sensing (such as water pipes) all require internet access, and there is considerable work worldwide going on in this area as a result. The next mobile telephone standard, 5G, is likely to be part of the solution—universal coverage and the ability to handle vast numbers of things connecting simultaneously is built into the technical specification. Achieving full coverage and (the perception of) good bandwidth at all times will almost certainly require a collaborative and holistic approach between telecommunications, broadband, wifi, and satellite providers. To get the best out of the connected future, individuals, communities, companies, and governments will all need to be more collaborative and interconnected. Products will need to be designed with connections in mind, services such as telecommunications will need to have considered the needs of transportation, healthcare, entertainment, agriculture, etc. when designing the performance of their service, and people will need to be willing to share data to give everyone better information.

As coverage improves, the increasing flow of data that follows will require an expanding infrastructure to support it. Data consumption is predicted to increase tenfold by 2020 according to CISCO, and there are mixed views on whether the power consumption increase associated with the data increase can be mitigated to some extent. According to the UK's *Independent* newspaper, data centers consumed about 3 percent of the world's energy in 2016, and the power consumption of data centers is expected to triple in the next decade. The proliferation of devices collecting, storing, and processing data for the IoT will be one driver for this growth, but it won't be the most significant, because many of the IoT sensors will be very low data—typical sensors will generate only a few hundred megabytes a year. Video uses about one gigabyte per hour and requires significant processing power as well as a lot of storage space. Video already accounts for most data on our networks, and entertainment, healthcare, lifestyle, and industrial uses are all likely to make increasing use of video. This is a serious energy issue, and data centers are already trying to improve the power efficiency of their plants. It is of course in their financial interests to do so. The big players such as Facebook, Google, and Apple are all pushing toward using fully renewable energy, building in cool climates to reduce cooling costs, as well as considering energy consumption in system architecture and operating procedures. Some predict that the use of video will have to be rationed or taxed in the future as one measure to contain the growth in energy consumption.

In addition to security, coverage, and energy, there are also societal issues emerging from technology advances, some of which have been touched on earlier. For example, the future of work: What will we all do when everything is smart and autonomous? What jobs will there be left to do by humans? And what will we learn in school when everything is changing so fast?

The future of work is an entire subject in its own right, but the types of jobs, how we educate for them, and how we create a society that highly values different types of work (such as caring) is a much-studied topic. There have been some social experiments by governments in countries around the world, such as Finland, Namibia, Canada, and the US, to see what impact a universal basic income or a shorter working day might have on society as a whole. Actually, the end of work has been predicted for centuries, but historically technology has always generated more jobs than it has destroyed—though it is too early to predict whether this trend will hold this time.

It is also likely that the IoT will help to create a sharing economy—a world where ownership is less important. We already no longer own our music or movies, we just pay for services to access them. It is likely that we will also be sharing cars, kitchens, and even pets in the future. In South Korea, the shared living room is already thriving. As apartments are relatively small and young adults often live with their parents, there is a healthy trade in social spaces to rent by the hour so that young people can socialize in home-like spaces. It's also possible that we will eat out more often, so the amount of space needed for a kitchen at home will be harder to justify. Future apartments may have many communal kitchens rather than a kitchen for every apartment. Parking garages in apartments built in California today are being designed with alternative future use such as gym or cinema in mind once shared cars are the norm. It's already possible (with appropriate checks) to rent a dog by the hour in New York as a walking companion. For people who struggle with the overall commitment of owning a pet (or occasionally need a cat to deal with a mouse problem), the option of buying a pet as a service could be appealing. The sharing economy raises issues around the future of manufacturing—in theory, we will need less manufacturing if everyone is sharing. Our appetite for consumption of products may actually decrease as our consumption of services increases.

Overall, despite the many significant challenges still to be overcome, the Cloud and the IoT are great enablers for major advances across society. The planet will require creative ideas and innovation—probably coming from "digital natives" who have grown up online—to see what could be possible and make the world a better place. The potential to improve quality of life, increase the productivity of companies, and tackle some of the big, global challenges based on the Cloud and the IoT is real and just waiting to be realized.

9

Cybersecurity

Alan Woodward

Contrary to the urban myth, the internet was not created as a network to withstand nuclear wars. The original internet (known as DARPANet because it was developed under the US Department of Defense's Advanced Research Projects Agency) was created to allow researchers to share resources freely. This philosophy continued when Tim Berners-Lee developed the Hypertext Mark-up Language (HTML), which became what we now call the World Wide Web. Then, in the mid-1990s, the internet was commercialized and everything changed. We quickly realized that if we were passing financial information across this newly available yet very public network, we had to have some way to keep it secure and private. Unfortunately, this new online world, or "cyberspace," had as its foundations technologies that were never designed to be secure. Cybercrime inevitably followed the money online to exploit this fundamental weakness, and thus cybersecurity was born.

Anyone who owns a personal computer will be familiar with the increasing onus of cybersecurity (or at least I hope they are): installing antivirus checkers and firewalls, practicing good password hygiene, and so on. As criminals become ever more ingenious, users have to spend more and more of their time exercising caution and keeping their security up to date. And, that's a problem. Even the most Spock-like of us exhibit human frailties that the criminals can exploit. Of the Seven Deadly Cyber Sins (apathy, curiosity, gullibility, courtesy, greed, diffidence, and thoughtlessness), apathy is probably the most dangerous, and thinking it can never happen to you is a certain path to ensuring it will. Solving the so-called PICNIC problem (Problem is In the Chair Not In the Computer) has caused computer scientists to wonder whether humans could simply be taken out of the loop.

Re-enter DARPA, which launched a competition to see if anyone could build an Artificial Intelligence–led system that could protect computers from cyberattacks, no human required. The first year of finals of the competition were held in 2016 in Las Vegas, where the world's hackers gather to exchange ideas and demonstrate their latest techniques at conferences such as Black Hat and DefCon.

The initial competition was quite modest in its ambitions. It gave the competitors several computer programs to analyze to see if they could spot whether the programs would crash if fed certain inputs, and then (the real trick) to modify so that they were no longer vulnerable to this kind of attack.

The automated systems were then pitched against humans in a Capture the Flag competition. The outcome? Well, that was a bit unclear: each system was good at some parts and not so good at others. However, one interesting thing that was happening was what is called "machine learning," and if there is *one* thing computers are very good at it is learning. As you provide them with more and more data they become better and better at recognizing patterns (such as problems that can be exploited by hackers). It's not surprising, then, that there are many research groups that are now turning their attention to how machine learning can be used in cybersecurity.

The fact that the AI systems in that competition weren't much better than humans at recognizing certain patterns is not that important. After all, it was a first attempt. What it did show was that AI could be the basis for cybersecurity in the future, which might already be looming on the horizon. At that Las Vegas conference, while the competition was under way, a company unveiled what it claimed was the first Artificial Intelligence "cognitive" antivirus system: DeepArmor.

DeepArmor may be addressing only one aspect of cybersecurity (antivirus), but it is the area that is most in need of automated help. This is because humans (or any process that has humans involved) simply cannot keep up with the torrent of new variants of viruses emerging. Nearly a million new threats appear each day, and while many are variants of existing viruses, each needs to be identified and the antivirus software told how to stop it. And this is only going to get worse, because the criminals themselves have also discovered that technology can help them evade antivirus software. They have developed viruses that metamorphose, just as biological viruses would do in the wild; even if you had a sample of the code that was launched it might have changed such that you would not recognize it after only a few infections. It is not altogether surprising, then, that those behind DeepArmor are releasing a piece of software that aims to mimic the human immune system: the aptly named Antigena.

This marks a radical shift in thinking about cybersecurity that will shape how we are defended in the future. The first epoch of cybersecurity on the internet has been very much about stopping attacks at the boundary. Rather like a castle, our systems—whether your laptop at home or your hospital's records—are set up to stop viruses, or any other form of attack, from entering. However, the hyperconnectivity in which we all take part means that these original ideas no longer work. So-called perimeter defense has been inadequate for some time. We actually want people to come into our castle. What we don't want is them leaving with the crown jewels. We have to have a means of allowing in potentially anonymous visitors, but we also need to be able to spot the suspicious characters.

One very recent evolution mimics the layered defense you see in many castles. We could allow people to wander around only certain parts of our castle, maybe even having different sets of people allowed access to inner portions of our cyber castle, hopefully keeping the crown jewels in the keep, the last redoubt of our security.

It turns out that machines are very good at watching behaviors and then correlating them with unwanted results. In fact, the more you let a machine watch the better it is at spotting those undesirable behaviors. Increase the dataset, and the results improve dramatically. So it seems our cybersecurity future may be sorted out: We can leave it all to the computers. But . . . employing Artificial Intelligence and relying upon it with no human intervention does raise some interesting questions, not least because it must initially be taught. It has produced an emerging discipline around aligning machine intelligence with human interests.

First there is the alarming possibility that the internet could become a form of electronic warzone with good AI and bad AI fighting it out. Just as we are not aware of the biological battles going on in our bodies every day, we may need to learn to live with ongoing digital infection that is constantly being fought by our "immune system." Most of the time we're unaware of the struggle but occasionally we "get an infection" and need antibiotics. But if the infection manages to outsmart the defending technology, the patient will effectively die. We may have to accept that in some cases we cannot "disinfect" our systems but instead have to wipe them and rebuild them with clean software. This is already starting to happen when computers are infected with ransomware, which leaves you with no option (assuming you're not tempted to pay the ransom) of rebuilding your system. The key, of course, is having a backup of your data, the only truly valuable part of your computer, and that requires humans to set it up.

The next concern is how far we allow AI to control our actions. Should we be allowed to override what the AI tells us to do? You might think the answer is an obvious "yes." Unfortunately, those

Seven Deadly Cyber Sins come into play again. Think of a scenario in which you attempt to visit a website, and the AI detects a problem and flashes up a warning to proceed no further. You might expect a button that allows you to ignore the warning, and research has shown that humans tend to ignore warnings, maybe because we are curious or apathetic, or simply because we don't like being told we cannot do something. Whatever your reason, if you proceed your system will most likely be infected. When you think about it that way around, it suggests the AI should have absolute control.

Research also suggests that the number of warnings that we would be presented with would be so large that we would begin to dismiss them anyway. This even has a name: Security Fatigue, which causes people to act recklessly. In order to avoid it, our future AI would have to decide which are the most important threats, judge how many warnings might be too many, and make some form of judgment as to when to present a warning, such that we might actually take notice. But of course AI does not exercise judgment, only humans do that; AI simply says yes or no.

All of which brings us to another concern. Which humans should be in the loop? For example, could the rules upon which the AI is operating be used to censor the sites you visit? It comes down to what is considered harmful. Cybersecurity has extended to trying to help the naive and innocent avoid visiting websites where they may encounter extreme material or be drawn into unwelcome behavior. If our future AI will also be "protecting" us from our own curiosity, then who determines the boundaries—or do we let the AI "learn" what is harmful and decide that way? Many understandably see this as the stuff of nightmares. Even without a despotic dictator or nanny state setting rules you disagree with, the AI might go further and decide that the vast majority of what we can find in cyberspace is harmful. The word "harmful" is remarkably ambiguous.

We have a decision to make over the next decade about whether we want AI protecting us in cyberspace or whether we want to rely on the current methods. It's not a simple decision, but one that will be made for us unless there is an active public debate. But, if we don't want AI and if the existing methods are increasingly ineffective, is there an alternative? Well, possibly.

Let's suppose we decide that a human has to be in the loop somewhere. What about situations where humans cannot be involved directly? What about the Internet of Things—that growing army of unattended smart devices that communicate without any intervention from us? Fridges, kettles, and toasters will all be networked to provide ever more convenient features. But while your networked toaster might not be worth hacking to plunder some valuable source of data, it can be forced to join an army of zombie devices in a botnet, ready to launch Distributed Denial of Service (DDoS) attacks where so much useless data is sent to a system that the system can longer deal with legitimate users. The size of the Internet of Things (IoT) will be staggering; millions more devices will be linked together than are networked today, and hackers will be able to subvert the spare capacity of these IoT devices. We face the somewhat ludicrous situation where a country could be brought to its knees by its own domestic products.

We could go back to basics, the basics of the technology upon which the internet was built. It would be possible to change some of the underlying technology such that wherever you go on the internet you are clearly identified, as are those with whom you interact. If you receive an email you could be sure that it came from whomever it claims to be from. If a website was launching malicious software, it could be easily traced and blocked. If an attacker probed your system you would know exactly who it was and where it came from. This all assumes that we change over to what is known as Internet

Protocol version 6 (IPv6). It has existed since the 1990s and is probably running on your computer today, but every attempt to spread its use has been disappointing. Most seem content to stick with the original IPv4 despite its inherent problems.

You would not need a new computer, but to make it secure does require (a) everyone to adopt it and (b) everyone being willing to have a digital identity. And there's the rub. Not everyone is willing to be identified online. People have become used to the internet as a place where they can roam free of the norms they may experience in their physical lives. We quite like being able to browse the mountain of websites in pseudo-anonymity. Some take this further and actively protect their anonymity through the use of software such as TOR (The Onion Router), which masks even their IPv4 address.

TOR was developed in part by the United States Navy with good intentions but has become infamous as the basis for running what are known as Hidden Services—websites that can only be reached by using the TOR browser and for which you cannot determine the physical location. This is what people refer to as the "Dark Web," and it hosts sites that offer to sell you everything from drugs to components for weapons of mass destruction. The Dark Web has now been joined by a new form of anonymous virtual currency, the best known of which is Bitcoin. These are "cryptocurrencies" and are intended to be the equivalent of online cash: untraceable and easily transportable. It's not surprising, then, that the European Police Office found that 40 percent of criminal-to-criminal money movements were done using this technology.

Equally unsurprising, law enforcement has made strenuous efforts to unmask those on the Dark Web. As with all technology, you can find technical means for undoing what the technology provided—in this case anonymity. However, it is a constant arms race, because as soon as a mechanism is found to unmask someone hiding in the Dark Web, the technology evolves so as to render the efforts of law enforcement useless. Even if some fundamental flaw in the way in which some piece of technology works is located, another emerges very quickly. For example, TOR depends to some extent on the volunteers running the nodes of this network protecting the anonymity of users passing through their systems. Once law enforcement began to exploit this fact to set up nodes from which they gained at least a partial vantage point of what was happening in the Dark Web, the technology was updated to prevent these so-called "bad onions" from undermining the anonymity of TOR users. Moreover, a whole new class of technology is emerging called Peer-to-Peer Dark Web, which bills itself as "The Invisible Internet"; unlike the TOR network, which is widely known about, P2P anonymizing networks are spread across the whole internet and thus blend into the background noise.

The same has happened with cryptocurrencies (which are a form of encrypted digital money). They typically use a technology called the blockchain, in which all transactions are public, cannot be forged or reneged upon, but are also designed to be completely anonymous. The classic policing technique is "follow the money," but even the most famous cryptocurrencies such as Bitcoin make this difficult by design, though techniques are entering service today that enable law enforcement agencies to have a chance of identifying where the virtual cash has gone. But, just as with TOR, as soon as the techniques were developed to determine who was using Bitcoin, new forms of virtual cash were deployed to render the new techniques useless. New cryptocurrencies such as ZeroCoin have emerged that are specifically intended to defeat the techniques developed by the police.

So, let's suppose IPv6 and its security add-ons become the default basis for the internet. There might actually be some law-abiding users, as well as criminals, who wish to remain anonymous.

Technologies like TOR might well persist (although they would require slightly different configurations), which seems to make introducing IPv6 pointless. However, the lack of success in introducing IPv6 and the security that can then be deployed with it has not stopped the enthusiasts. It will catch on, eventually. We might end up with a situation where we have a two-tier internet: those who are happy to be identified and potentially tracked and those who wish to remain outside this system. In some countries it will not be a choice, but where it is you could envision a walled garden within which you can operate free from fear of attack. But, if you viewed this walled garden as more of a gilded cage, you would be free to leave the walled garden and enter the remainder of the internet—a part of cyberspace rather like the Wild West. The choice would be yours, but you might find that if you have actively chosen to leave the sanctuary of the walled garden you might not be allowed back in. After all, if you live in a sterile zone you cannot leave and reenter without a whole series of checks being done to ensure you haven't brought something nasty home with you.

Hackers would see those wandering in this protected space as juicy targets. It would be inevitable that hackers would find a way of surreptitiously entering the safe space and exploiting the unsuspecting. Today we are familiar with the use of social engineering by hackers—the use of a known email address or piece of personal information to make us click on a link, for example—but if they were completely unexpected, social engineering attacks would have a much higher impact. It would be a very much more difficult exercise inside our walled garden, but it would happen.

The Internet of Things devices that would be running as part of our walled garden could prove to be the weak point. If you look back at the most successful bank robbers, they do not break down the front door but sneak in via the sewers so that they are long gone by the time anyone knows they have emptied the vaults. No cyberspace would be truly safe, but it would be a lot safer than the cyberspace we know today. The question is, which side of the wall would you want to be on?

It has been suggested that our future cyberspace may need two different forms of policing. Inside our safe space we would still need some policing, even if it is only to settle commercial disputes—more akin to the neighborhood cop on the beat who knows everyone and steps in to have a quiet word when needed. Outside the walled garden you would need something more like the marshals of the old Wild West. These cyber marshals would be specialists with a mandate to seek out and contain criminals online. Such specialist policemen have existed in other domains before. For example, when the subway arrived in New York, new forms of crime followed, and eventually the Transit Police Department was formed. When terrorism made its way onto our airlines, air marshals were deployed to watch out for those that might do others harm. And because the internet is a global phenomenon we may even see United Nations cyber blue helmets deployed to keep the peace outside the safe zone. A key issue will be who it is that gives these cyber marshals their mandate.

Whether inside or outside our potential cyber safe zone, a big question would be which laws apply. At the moment, legislation typically follows national boundaries, but cyber criminals do not. If I send millions of phishing emails and commit fraud from one country aimed at recipients based in another country, there is far less risk of me being pursued, let alone prosecuted, than if I walk into a bank with a sawn-off shotgun. And the returns are much higher: cybercrime can deliver you a tenfold return on investment with very limited risk. No surprise, then, that online felonies are now the single biggest form of crime, and yet we still have no single cyber law enforcement agency.

To tackle this we could try to negotiate international treaties just as we did to govern telecommunications or air travel, but cyberspace is a great deal more complex. It took decades to agree whose law governs which situation involving air travel, telegraph, and telephone, and these all had relatively few dimensions. For a start, cybercrime assumes that everyone views certain acts to be criminal, which is not so. For example, many nations around the world have agreements that they will protect the copyright of authors, musicians, filmmakers, and so on across their borders in the same way that they are protected in their country of origin. However, some countries, notably from the former Soviet Union, Eastern Europe, and Asia, have proven more reluctant to enforce these copyright rules than others. Not surprisingly, it is now these few countries that are responsible for most of the distribution of illegally copied material.

Maybe we would end up with a situation where the cyber marshals were the law in the lawless portions of the internet, or even with cyber vigilantes maintaining law and order. This is not a new phenomenon, but at least law enforcement agencies currently apply the law equally, so if these characters step over the line they are as culpable as anyone else. History has taught us that vigilantism in a completely lawless environment can lead to situations where the cure is worse than the disease.

So, we are faced with an interesting dilemma: whether we want to have the black-and-white vision of AI security or the fallible human judgment of a cyber marshal. There are no easy decisions for our future security in cyberspace. The only thing you can guarantee is that criminals will fill any vacuum left by law enforcement, and we collectively need to decide if we want to be a collection of individuals looking out for ourselves or if we want to hand over our security to someone, or something, else. Gort (the peacekeeping robot in the film *The Day the Earth Stood Still*) or Wyatt Earp—you decide.

10

Artificial intelligence

Margaret A. Boden

It's impossible to say just where artificial intelligence (AI) will be in the future. But it's easy to say where it won't be: It won't be hiding under a stone. Instead, it will be ubiquitous and inescapable.

AI is already active in many aspects of society. It's at the heart of every internet search and every app. It's in every GPS query, every video game and Hollywood animation, every bank and insurance company and hospital—and, of course, every smart watch and driverless car.

In the future, it will be found everywhere: law courts, offices, old people's homes ... even marriage guidance counseling. Mars robots will have countless cousins working on warehouse floors. The Internet of Things will interconnect wearable computers (monitoring our location, activity levels, and blood pressure) with a host of gadgets in our homes, offices, streets, and restaurants. It won't be Big Brother watching you, but rather trillions of little brothers—and all talking to each other nonstop.

These developments will happen quite soon—in fact, most have already started. Within a couple of decades, countless such examples will be shaping our lives. Industrial societies will be deeply dependent on them. And developing countries will be affected, too, as AI-based medical or agricultural advice (for instance) is made available to people living miles away from modern hospitals or agri-scientists.

One recent advance that has made all this possible is machine learning using Big Data—huge caches of information that can be analyzed for patterns and trends in human behavior. This AI technology—known as "deep learning"—has been known about in theory for over a quarter of a century but couldn't be put into practice because computers weren't sufficiently powerful. Over the past few years, however, computer power, and computer storage, have increased enough to enable today's machines (doing a million billion calculations per second) to learn from enormous data-banks holding billions of items.

This type of machine learning can find patterns, on various levels of detail, in huge collections of data. These include (for instance) the current speed and location of every car and bus on a city's roads, with the current state of every traffic light; or the medical records from every regional or national hospital, detailing every patient's symptoms, drug dosages, and results.

Such AI systems aren't programmed in the traditional sense of "do this, then do that." Instead, they consist of multilayer neural networks, the output of one layer being fed in as input to the next. Each layer contains many thousands of units that communicate with each other until they "settle" into a state representing a stable pattern found in the data. Often, these patterns are new and unexpected—even unsuspected by the humans running the systems.

In 2016, an application of this technology (developed by Google DeepMind) learned to play the game Go well enough to beat the world champion—a feat much harder than beating the world chess champion (done by IBM's Deep Blue in 1997). But while this was just a game, impressive as a demonstration, it was of no practical use. However, deep learning is already being used by governments, and by resource-rich companies around the world. As the technology becomes cheaper, and so more widely available, it will spread into virtually every area of society.

Perhaps you're thinking "Go, today. Anything you want, tomorrow!" However, it's not quite like that. Today's learning systems are hugely powerful, but they're not well understood. In the jargon, they are "black boxes": systems that have measurable inputs and outputs, but whose internal workings remain obscure. Their designers/programmers don't really know how they work, so can't reliably predict what they are going to do next. The designers are well aware of this problem, of course, and massive efforts—and budgets—are being devoted to it. But no one knows when, or even whether, these black boxes will become manageably grey (by which I mean somewhere between black and white—in which we understand exactly how they do what they do), and it is probably just a pipe dream.

So we have to be careful with our futurology. Some things are very much more difficult than most people believe. One such example is artificial *general* intelligence (AGI). This was the goal of most of the AI pioneers in the 1950s—and of Alan Turing, even before that. They hoped to develop AI systems that would match the generality and flexibility of human intelligence. One and the same program—doubtless, a very complex one—would be able to solve problems of many different types. It would use—and intelligently integrate—language, vision, hearing, learning, and creativity. Add motor behavior, too, if it was controlling a robot. In other words, this hoped-for program would be very different from the single-minded, super-specialist, programs and apps that we're familiar with today.

Initial hopes for AGI were high. One program of 1959 was even named "the *General* Problem Solver," because—in principle—it could solve *any* problem that could be stated in terms of hierarchies of goals and sub-goals. (Actually, stating a given problem in such terms was left to the programmer, and was the hardest part of the exercise.)

One of the problems it tackled successfully was the missionaries-and-cannibals puzzle: *Three missionaries and three cannibals stand on one side of a river with a boat able to carry just two people; how can all six cross the river, without ever having missionaries outnumbered by cannibals?* That success wasn't trivial, because the puzzle has a catch in it. (Try it yourself, using coins.) It was used as a benchmark for AI at that time, so it's not surprising that people got very excited when it was solved by a machine.

A lot of effort went into improving, and widening, such systems. But it proved too difficult. By the end of the 1970s, most AI researchers had turned to much narrower tasks, such as "expert systems" designed to diagnose and/or prescribe drugs for a particular disease. The motto used in the field was "putting world knowledge into AI programs" because they were being provided with factual knowledge (about medical diagnoses, for example).

Today's expert systems (a term that is no longer fashionable) comprise many thousands of examples, used for both complex and trivial tasks. These range from oil exploration, through machine translation and face recognition, to finding the nearest Indian restaurant. Most AI research in the twenty-first century concerns specialist systems such as those. But not all of it, for AGI is now a respectable goal again. (The LIDA model of consciousness, mentioned below, is one example.) However, no current AGI system is truly "general" in the sense that Alan Turing envisioned.

Fans of IBM's Watson program might dispute that, however. They would say that AGI has already been mastered. After all, in 2011 this program beat the two human champions at the general knowledge game of Jeopardy! Contestants in that game aren't asked straightforward questions. Instead, they're given a clue and asked what the appropriate question might be. (Not "What is the capital of France?," but perhaps "A capital occupied by a hotel heiress.") Such a problem requires lateral thinking—or at least wide-ranging associations.

Watson's many successes at Jeopardy! are a genuine achievement, of which its designers can be rightly proud. However, Watson's intelligence is very unlike ours. On one occasion, for instance, the system correctly focused on a particular athlete's leg, but didn't realize that the crucial fact in its stored database was that this person had a leg missing. The programmers have now flagged the importance of the word "missing" so that this mistake won't occur again. But a human, knowing the relevance of a missing leg to athletic prowess (and to everyday life), wouldn't have made the mistake in the first place. This issue of human relevance, or human life, is the major bugbear facing all AGI systems—put simply, computers don't know what it is like to be us.

In other words, some ambitious AI/AGI tasks may be so difficult that, even if they're achievable *in principle* (the human brain doesn't work by magic, after all), they're forever impossible *in practice*.

This is denied by people who believe in "the Singularity"—the imagined point that some AI scientists say is only twenty years ahead and at which AI will equal and then surpass human intelligence as machines rapidly, and intelligently, improve themselves. (The forecasts vary: Many AI scientists foresee the Singularity happening by the end of the century.) All our major problems, say some Singularity-believers, will be solved. War, poverty, famine, illness, even personal death: all banished.

This view is hugely controversial. Is it really credible that some non-human system could solve the political crises in the Middle East, for instance? Admittedly, we humans haven't made a good job of it so far. But that AI could ever do so is a belief that requires a leap of technological faith that's too far for many people (myself included); the political sensitivities and historical background involved are far too complex and subtle for AI to cope with.

Another futuristic scenario foreseen by (some) Singularity-believers is one in which "the robots take over," with horrendous results for humanity. In this scenario, the super-intelligent AIs will follow their own goals relentlessly, perhaps greatly to our detriment. They needn't (although they might) actually try to harm human beings. But, much as most humans are indifferent to the fate of ants, the super-intelligent AIs may harm or even destroy us if we get in their way. A future AI designed to manufacture paper clips, for instance, might grind up human bodies so as to extract the metallic atoms—the iron in our blood, for instance—that are usable in the production of paper clips.

In brief, people disagree vehemently not only about whether the Singularity will ever happen, but also about whether that would be a Good Thing or a Bad Thing. In either case, there would be plenty of excitement ahead. But one doesn't have to believe in the Singularity to predict an exciting future for AI. It's a safe bet that AI will go very much further than it's gone today. But just how far *could* it go? Could the AI of the future pass the famous Turing Test, for instance? In 1950, Turing foresaw a time when someone could converse with an AI program for up to five minutes without (for 30 percent of the time) being able to tell whether it was a computer or a person. The test hasn't been passed yet—although people have often been fooled if they weren't warned that they might be talking to a computer.

There is a Turing Test competition (shown live on the Web) held every year at Bletchley Park, where Turing helped break the Germans' Enigma code in the Second World War. The contestants hope to win the Loebner prize: $2,000 for the "best" entry each year—with $25,000 promised for "the first computer whose responses are indistinguishable from a human's," and $100,000 for a human-seeming system possessing audio-visual capabilities as well as language.

As yet, the $25,000 remains unclaimed. No program has fooled the judges for the required 30 percent of the time. Admittedly, the

Loebner competition was won in 2014 by a program that fooled 33 percent of the interrogators into thinking that it was a human being. However, the human being they had in mind (i.e., the choice they picked, out of the several offered to them by the test organizers) was a thirteen-year-old Ukrainian boy. In other words, the (English) language used by the computer was far from perfect. Its errors and clumsiness were forgiven by its human interlocutors, much as we naturally make allowances for the mistakes of foreigners, especially children, who aren't speaking in their native tongue.

The obvious question, here, is "So what?" Suppose that an AI system did pass the Turing Test one day—or perhaps what's called the Total Turing Test, which would involve a robot with human-like sensorimotor behavior. What would that prove? Would it mean that some computers can really *think*? Would it show that some can actually be *conscious*?

"Consciousness" is a slippery concept. But one can distinguish between *functional* and *phenomenal* consciousness. The first of these categories covers a variety of psychological distinctions. These include: awake/asleep, deliberate/unthinking, attentive/inattentive, accessible/inaccessible, reportable/non-reportable, self-reflective/unexamined, and so on.

Those contrasts are functional ones. There are good reasons to believe that they can be understood in information-processing terms, and therefore modeled in computers. (The most interesting model of machine consciousness at present is LIDA, which addresses the distinctions just listed—and is based on a widely accepted theory of brain functioning.) So, some future AI system might well be conscious in that sense. A robot passing the Turing Test, for example, could properly be said to plan and to think.

Phenomenal consciousness, or *qualia* (such as pain or sensations of the color red), seems to be very different. Its very existence, in a basically material universe, is a notorious metaphysical puzzle. Explaining phenomenal consciousness is sometimes called "the hard problem," because solving it is so much more difficult than explaining how functional consciousness is possible.

Various highly speculative, seemingly crazy, solutions have been suggested to explain qualia. Some of these treat phenomenal consciousness as an irreducible property of the universe, analogous to mass or charge. (This hardly solves the problem, of course!) Others appeal to quantum physics, using one unfathomable mystery to try to solve another. And many people simply throw up their hands in despair. One prominent philosopher has claimed that not only has nobody "the slightest idea how anything material could be conscious," but also that "nobody even knows *what it would be like* to have the slightest idea how anything material could be conscious." In short, the topic is a philosophical morass.

If phenomenal consciousness in human beings is such a deep mystery, we aren't in a good place to judge whether or not some future AI system might have it. Granted, the suggestion may seem ridiculous to you. It certainly seems ridiculous to me. But that's an unargued intuition, not a carefully reasoned conclusion.

Most AI scientists—LIDA's programmers, for instance—ignore phenomenal consciousness. They see it as just too difficult. But a few AI-influenced philosophers analyze it in information-processing terms.

Qualia, on this view, are internal computational states—within the program, or "virtual machine"—that are implemented in the brain. These states can have causal effects on behavior (e.g., involuntary facial expressions). They can also cause changes in other aspects of the mind's information processing (e.g., planning for revenge, if someone causes one pain). They can exist only in computational systems of significant structural complexity. (So snails, for instance, probably can't have them.) They can be accessed only by some other parts of the particular system concerned, hence their "privacy." Moreover, they cannot always be described—by higher, self-monitoring, levels of the mind—in verbal terms, hence their "ineffability." In a nutshell: "You can't experience *my* sensations of the color red. And I can't *describe* them fully to you (or to myself), either."

If such an account of qualia is correct, then some future AI systems could have phenomenal consciousness. On this question, therefore, the jury is still out.

The jury is still out, too, on the question of whether AI *is*—or, more to the point, *will be*—a Good Thing. The Singularity enthusiasts mentioned above answer with a resounding "yes!" Even if one doubts (as I do) that the Singularity will ever happen, one must admit that today's AI is helpful for many purposes—and that this will be even more true in the future. So much is certain.

To that extent, then, the "Good Thing" label seems about right.

Nevertheless, there are worrying aspects to AI's future—some of which we should be thinking about seriously already. One obvious example is the threat to employment—not only the number of jobs (professional as well as menial) but their down-skilling as the more routine aspects are taken over by machines. Ideally, jobs will be restructured so that humans will work alongside machines; the people will do what only people can do, leaving the rest to be done by AI. Even so, many of the people who would have been doing the boring stuff will find employment difficult. To be sure, some new jobs will be created. But these will probably require a level of education way beyond what's available to most people, even in the most advanced societies.

Some might say that employment doesn't matter. Wage packets, they suggest, will be substituted by some "universal basic income," provided to every citizen as a right. (Experimental versions are being run or planned in various countries.) But that is problematic. Where is the tax base to support such societal largesse? And what will people do with so much leisure time? Social psychologists have shown that having a job, even a menial one, provides very much more than money.

Another obvious example, or set of examples, concerns the military. Drones are already dropping bombs, but the targets are selected with human input. A nightmare future would see fully autonomous drones, with the targets selected by machines. This is discussed in the chapter by Noel Sharkey.

In addition, there's a large group of examples of AIs being applied in activities that today involve empathetic face-to-face communication between people. Significant sums of money are being put into research on "computer caregivers," "computer companions," and "robot nannies."

If a robot nanny is merely a system that monitors the baby's/infant's crying or sleeping patterns, and alerts the human caregiver appropriately, that's fine. But if it is also a natural-language processing system that's supposed to entertain and educate the infant, then that's not so good. Even watching *Bambi* with the child might cause huge problems: What can the robot nanny say when Bambi's mother is shot?

Similarly, if the old people's homes of the future employ AI systems (on-screen or robots) to do mundane tasks for the elderly users, that's fine. But getting into personal conversations with them, and arousing emotion-laden memories, could do very much more harm than good. (Of course, in an ideal world, such AI applications wouldn't even be considered. The professional human caregivers would be paid, and respected, much better than they are today—and the old person's friends and relatives would take the trouble to visit. But do you want to hold your breath?)

A number of groups around the world are thinking about these varied issues now and trying to find practical ways of preventing, ameliorating, and regulating them. One, for instance, is an international committee, chaired by a roboticist, that tracks—and advises on—the development of robots (such as drones) for the military. Others include groups of AI professionals and governmental policymakers who monitor the use of Big Data in various fields, looking out for issues surrounding violation of personal privacy, for instance.

In sum, there are major changes ahead. Many AI developments will improve our lot significantly. But some will have unintended (though sometimes foreseeable) consequences that threaten important aspects of human lives. AI research shouldn't be given a completely free rein.

11

Quantum computing

Winfried K. Hensinger

When I grew up in the early 1980s, conventional computers had only recently become commonplace. In fact, I learned my typewriting skills on a mechanical typewriter, and life in general was quite different—no electronic ticket machines at train stations, no internet, no smartphones. In fact, common tasks did not rely on the existence of computers and many things we take for granted today simply did not exist. Conventional computers today have an impact on nearly all aspects of our lives, and we cannot imagine a life without them. Scientists and politicians have coined a term for a technology that changes all of our lives: They call it "disruptive." The term is used because of the potential of a technology to transform our lives in a step-changing, abrupt, and overarching way. In the next ten to twenty years, I believe, we will see the emergence of another disruptive technology: quantum computing.

Let's be clear about something first: quantum computers are not simply very fast computers. In fact, they have very little to do with conventional computers, and it is unlikely they will be used to tackle tasks for which we currently use conventional computers. In contrast, quantum computers may enable us to solve certain problems that we never even thought could be solved, a class of problems that might take even the fastest supercomputer billions of years to crack. Quantum computers will likely provide us with entirely new capabilities and therefore change our lives in very unexpected ways. To appreciate some of these capabilities, and to imagine the ways in which they will change our lives, it is best to start at the beginning and to describe what the working principle of a quantum computer actually is. In order to give you an idea, let me start, if I may, with a small introduction to quantum physics.

In short, quantum physics is a theory that explains the world around us. However, it is a rather strange theory. For a start, it predicts that something can be at two different places at the same time. Yes, you heard that right: Quantum physics, in principle, allows me to sit on my desk in Brighton in the UK to write this chapter while simultaneously going for a swim on a beach in Florida. Unfortunately, this does not happen with very large objects such as people (I certainly wish right now that it did). However, it is quite regularly observed in the laboratory when studying the behavior of individual atoms. Indeed, an atom can be at two separate locations at once. This phenomenon is referred to as "superposition." Physicists have been stunned by this strange prediction of quantum physics and have carried out numerous experiments trying to disprove it. However, experiment after experiment has shown that this can and does indeed happen.

Let me tell you about an example of quantum weirdness from my own scientific career. A quantum physicist by the name of Gerard Milburn had predicted that it should be possible to make an atom move forward and backward simultaneously. To understand the idea behind this, imagine you are trying to get out of a tight parking space in your car. Rather than moving forward and hitting the car in front of you then reversing to bump the car behind you, you are hitting both the car in front and the car behind—at the same time! This idea utterly fascinated me when I was young, so it was an experiment I wanted to try when I started my scientific career. After a lot of hard work and many long nights in the laboratory over a period of about three years, we finally observed just that—an atom moving forward and backward at the same time—an example of quantum superposition.

And as if superposition were not strange enough, things get even wilder! There is another phenomenon, called "entanglement," that's even weirder. In fact, the only correct way to explain entanglement is via mathematical equations. However, let's try to use a much simplified description. It is possible to entangle two quantum objects, such as two atoms, so that if I were to do something to one of these atoms this would instantly influence the other one, even if it is located far away and there is no possibility for it to communicate with the first atom. Einstein was very uncomfortable with this feature of quantum physics. He called it spooky and proposed experiments that would disprove it. Physicists have been carrying out such experiments for the past sixty years, refining them and homing in on every kind of possible loophole they could imagine. However, the results are always the same: quantum physics seems to be correct, and strange phenomena such as superposition and entanglement do indeed happen.

While some physicists are still trying to understand the strangeness of quantum physics, many others have accepted it and have sought a new challenge. They ask the question: Is it possible to harness such effects to build entirely new technologies that rely on the strange predictions of quantum physics? There are numerous ideas, such as the possibility of a new generation of sensors that may detect electromagnetic fields with unprecedented accuracy, or even measure gravity itself (in order to detect pipes underground, for example). Other applications include quantum cryptography, which would allow us to communicate securely, safe in the knowledge that the laws of physics themselves are protecting the communication from ever being eavesdropped upon.

All of these are tremendously exciting technologies with the potential to have a significant impact on our lives. However, in my opinion the most step-changing (and hardest to realize) technology is quantum computing. So let me explain why this would be such a novel and powerful machine.

Quantum mechanics explains how atoms behave and interact with each other to give rise to all the properties of matter in the universe, such as an object's color, its strength, and the way it conducts heat and electricity. It also explains how the atoms in our own bodies behave, giving rise ultimately to how we see, smell, and generally experience the world around us. It is an extraordinarily powerful theory.

But there is one big problem: Predicting and calculating quantum mechanical processes using conventional computers is very difficult. In fact, solving nearly any quantum-mechanical problem exactly is intractable even for the most powerful conventional computers, as quantum physics calculations require extensive computational powers. It would take conventional computers billions of years to find the solution for many of the really interesting problems to be computed. One could maybe summarize the work of the majority of scientists around the world today as creating highly simplified models of quantum processes in such a way that they can be computed on a conventional computer. However, compared with solving the exact quantum problem, such inevitably simplified calculations cannot always model these processes accurately. This in turn means that we are missing out on tremendous opportunities that would potentially enable us to create new pharmaceuticals, make new materials, understand protein folding, and control and understand processes in biology, to name just a few applications. A quantum computer may be able to solve such problems because it relies directly on the stranger features of quantum mechanics, and as such is capable of accurately simulating complicated systems that are based on quantum physics! We are only just starting to realize some of the groundbreaking consequences the use of quantum computers will have. Do we yet know the most important breakthroughs that will result from using quantum computers in understanding and controlling other physical systems? Unfortunately, we can only speculate. However, since all physical systems and their various properties can be explained by quantum physics, it does not take a PhD in physics to realize that the opportunities may well be endless. In fact, using quantum computers to understand reality itself constitutes an unprecedented approach and thus may well drastically change our understanding of the universe, and even life itself.

This should already sound wildly exciting. However, there is a second application of quantum computers that may have an equally dramatic impact on all our lives. To understand this, it is best to start with some explanation of conventional computing. The computational power of conventional computers has been steadily growing over the last thirty years, approximately doubling every eighteen months, a relationship known as Moore's law. This has been achieved by miniaturizing the transistors making up the processor inside a computer. It sounds like an impressive continuous improvement in computing power. However, certain problems are so complex that the best computers available would still require far too long to calculate the correct answer. Examples of such problems would be to accurately predict the weather, or the smartest possible investment on the stock market in order to maximize return. Another application might be to calculate the fastest route for a courier company to make multiple deliveries in one trip. As the size of such problems increases (such as the number of required deliveries or the range of weather forecasting), so does the number of parameters required to model such problems. Therefore, conventional computers struggle to finish the calculation due to the enormous number of computational resources required.

Quantum computers, however, may hold the answer. In a conventional computer (which we refer to as a classical computer to distinguish it from a quantum one) information is encoded in binary bits, with each bit being either zero or one. A string of bits then represents the information to be computed by the processor—for example, a number. Let's take the example of two bits. We could write into our computer any combination of two bits, for example, we could choose 01 (the binary number "one") and 10 (the binary number "two"). These two numbers are then sent in the processor and added together. Once an answer has been computed we would write the next number into our 2-bit memory, in this case, 11 (which represents the number "three"), which is then used in a subsequent operation to give the next answer. What defines the operation of a classical computer is that these calculations are carried out sequentially, one after the other. However, a quantum computer can make use of the principle of superposition. This means that two quantum bits can hold all two-bit combinations, namely the four numbers 0, 1, 2, 3 (written in binary as 00, 01, 10, and 11) simultaneously! And inside the quantum processor all the calculations are carried out at the same time, too! In fact, in one of the interpretations of quantum physics, it is argued that these calculations are actually carried out in parallel universes. Comparing two bits with two quantum bits, the difference between a classical and quantum computer does not appear very significant (e.g., four numbers are computed in a quantum computer instead of only one in a conventional computer). However, as we go to a larger number of bits this difference rapidly increases. Ten quantum bits can already hold 1,024 different numbers simultaneously, while one hundred quantum bits can hold 1267650600228230000000000000000 different numbers—all at the same time. While a quantum computer can compute all of them simultaneously, a classical computer would have to compute one number after the other. Now it becomes clear why quantum computers are so powerful.

There is, however, an important caveat. While a quantum computer can do all these calculations at once, it is not possible to read out all the answers; in fact, one can only read out one of the answers. Nevertheless, there is a way to overcome this limitation. It is possible to make sure that this one answer depends on all the calculations—actually making use of all of them. So, in order to make use of the opportunities of quantum computers, we need to focus on problems where we are interested in one answer that depends on many individual calculations or operations. Searching through a database is such a problem. Imagine you are trying to find the name associated with a particular phone number using a phone book. On average, you have to look at half the entries before you find the name you want. Sometimes you might get lucky and find it quickly, other times it will be closer to the end, but on average—after many such searches for different names—you will have covered half the phone book. This is exactly the type of problem a quantum computer could really excel at, and indeed search algorithms are among the applications where quantum computers have been shown to provide significant potential speedup compared with classical computers. The creation of algorithms suitable for use in a quantum computer is an emerging field of study in itself, and we are only at the beginning. Encouragingly, many very potent algorithms have already been found (see the NIST Quantum Algorithm Zoo website for a list of currently known quantum algorithms).

So how do you go about actually building a quantum computer? Unfortunately, this is unbelievably difficult, and for decades scientists felt it might not even be possible. However, with numerous breakthroughs in recent years, this view has changed. I think it is fair to say that from what we know it is possible to build a quantum computer with current technology. However, considering the required engineering, it might be as complicated and ambitious as manned space travel to Mars. But let's have a closer look at what is actually required.

The most important ingredient is a physical system that exhibits quantum effects, since we require the quantum phenomena of superposition and entanglement for the machine to work. The good news is that this allows us some flexibility, as any physical system in principle exhibits quantum effects. Indeed, when quantum computers were first envisioned, scientists made proposals for numerous physical systems that could be used. The list includes silicon wafers with atoms of other elements forming impurities on the surface that then form individual quantum bits, or individual electrons floating on helium, or charged atoms (ions), superconducting circuits, photons, and numerous other systems. Countless ideas have been explored.

While all these systems certainly exhibit quantum effects, the hard part is having the ability to fully control them—to make them happen on demand. While physicists have great experience in observing the strange phenomena predicted by quantum mechanics, controlling them is extremely hard. Part of the reason is that any unwanted interaction will immediately destroy these effects (this is in fact the reason we do not see larger objects, such as people, being at two different places at the same time). While many physical systems are still under investigation and may well provide a fantastic architecture for quantum computers in the future, two leading candidates have emerged where impressive progress has been made—so much so that we now believe it is indeed possible to construct a large-scale quantum computer.

One of the contenders uses a quantum phenomenon called superconductivity, whereby for such a quantum computer to work it has to be cooled to close to absolute zero (–460°F or –273°C). This can be readily done with just a few qubits, but the engineering becomes considerably more challenging when billions of qubits are involved.

The other contender, and the physical system with the best specifications achieved so far, is trapped ions (ions are charged atoms), which can operate at room temperature or may only require relatively "modest" cooling (to –321°F), the temperature at which nitrogen gas liquefies). Only a few months ago, my group at the University of Sussex, with the help of some extraordinary scientists from Google, Aarhus University, RIKEN in Japan, and Siegen University, published the very first construction plan to build a large-scale quantum computer using trapped ions. And we are now in the process of actually building such a device at the University of Sussex.

So let's discuss how a quantum computer can be built by making use of trapped ions. Each ion forms one quantum bit. The ions are trapped in a very good vacuum to make sure that the atoms representing the quantum bits do not collide or interact with any other atoms located somewhere inside the system. The ions are held in place using electric fields that are emitted from electrodes located on specially designed microchips. The electrodes on such microchips form arrays of cross-junctions, giving rise to a large labyrinth, similar in nature to a game of PAC-MAN, along which ions can be transported by changing voltages on these electrodes. Ions are moved across the surface going from memory regions to quantum gate regions where logical quantum gates (the computations themselves) are carried out. Traditionally, such quantum gates would have to be realized utilizing individual pairs of laser beams that need to be aligned with micrometer accuracy. Nearly as many pairs of laser beams as there are quantum bits would be required for the calculation—potentially billions of them. While applying this technique to a handful of ions has been demonstrated, imagine how a quantum computer would have to be engineered to host billions of such pairs of laser beams in order to cater for billions of ions.

Fortunately, we recently invented a new approach where all these laser beams can be replaced with voltages applied to a microchip. This drastically reduces the difficulty of building a large-scale quantum computer. In the blueprint to construct a large-scale quantum computer, we aimed to provide a good overview of all the engineering tasks that would be required. In order not to rely on future breakthroughs in physics, and as such to be able to say that we are now able to construct a large-scale quantum computer, the machine we envision will not be as optimal is it could be; it will be large (the size of a building or even maybe a soccer field), very expensive, and may still take ten to fifteen years to construct. However, there are no obvious fundamental physics-related obstacles stopping us from building such a machine. From the above it becomes obvious that we are nowhere close to making compact quantum computers for use in the home. However, that probably does not really matter. After all, the first conventional computers were not small, either; they, too, filled a whole building. In the age of Cloud computing, it seems a lot more natural to have the quantum computer at a central location with users accessing it remotely in order to run a calculation.

So where are we right now with the realization of commercial quantum computing? A Canadian company, D-Wave, took a bold step and marketed their technology as a quantum computer. This has been viewed with skepticism by many physicists. Recent studies seem to confirm that certain quantum processes do indeed play a role in the working of the D-Wave machine. However, I have not seen any evidence that these machines will ever have the capability to become universal quantum computers—in the sense that they will have the ability to carry out the full range of applications envisioned for a general quantum computer. This is because the quantum bits used inside the D-Wave machine do not have the same capabilities as, for example, the quantum bits within an ion trap quantum computer.

As such, D-Wave machines could be considered as special purpose machines without the flexibility to evolve to become a universal quantum computer. While this does not sound so impressive, there may well be some interesting applications for this machine.

Building a large-scale universal quantum computer is considered to be one of the holy grails of science. At the same time, many major companies have taken an interest in developing quantum computing, realizing that competence with this technology may well be critical for their survival—examples include IBM, Google, and Microsoft, as well as some startup ventures such as IonQ. And there are a number of universities working on the construction and commercialization of quantum computers. It is fair to say that most of the known fundamental roadblocks to building a quantum computer have now been removed. However, the engineering is still pretty daring, so the first large-scale machines are still a decade or two away, and we will undoubtedly see a range of technological solutions in the quest to build a large-scale machine. The opportunities that this new technology offers may provide for an entirely different world, and we are very lucky to be part of a generation that will likely experience some of the wonders that will come from it.

MAKING THE FUTURE

Engineering, transportation, and energy

12

Smart materials

Anna Ploszajski

Imagine what life would be like if your possessions could sense, react, move, adapt, morph, and repair themselves completely independently. In the future, this will be a reality; solid objects will carry out useful functions for us without any need for human interaction, not by using robotics or electronics, but by being made of "smart materials." These are solids with properties—such as color, shape, or magnetism—that change autonomously in response to stimuli such as light, temperature, applied force, or moisture. The scope of this topic is vast. Within our lifetimes, we will see smart materials everywhere: on color-changing roofs to regulate the temperature of buildings, as wearable displays, as the fabric of human-like robots, or even as self-opening cans of baked beans.

Smart materials are not new. Indeed, nature got there before us with pine cones that close when it rains and plants that grow toward the light. We've even used smart materials throughout our own history: four and a half thousand years ago the pyramids of Giza were plastered with a self-healing lime mortar. But the first time a smart material was recognized as such by scientists was in 1880. Its discovery was made by brothers Pierre and Jacques Curie, husband and brother-in-law, respectively, of Marie Curie. They found that compressing a crystal of quartz, a common transparent mineral found in granite, caused an electric voltage to appear across it. A year later, they proved that the effect was reversible—applying a voltage across the crystal caused it to physically deform. They dubbed the effect *piezoelectricity*, from the Greek *piezo*, to squeeze, and *elektron*, meaning amber, an ancient source of electric charge. Piezoelectric crystals were first applied during the First World War in sonar detection devices; today they are used in many applications, from cigarette lighters and microphones to clocks and ultrasound imaging.

The Curie brothers' breakthrough inspired materials scientists, engineers, and inventors to rethink their approach to materials design, which led to the discovery of a whole catalog of new smart materials. Today there are millions of patented inventions that use them. Broadly, their functions fall into six categories—color changing, sensing, moving, heating/cooling, self-healing, and phase changing (freezing and melting). And smart materials are by no means limited to the realms of science fiction or the laboratory—most people will be familiar with a few already, such as the photochromic sunglasses that darken in reaction to sunlight, or thermochromic mugs that change color with hot coffee inside.

A ride on a futuristic smart bicycle will be a carefree experience, since smart materials will make potholes, punctures, and paint scratches a thing of the past. You'll be able to cycle whatever the weather, thanks to clothing that quickly adapts to body temperature and rain. If you get caught out after dark, roads will be lit sustainably by the weight of passing vehicles, and if you happen to fall off, any ripped smart clothing will self-repair on the roadside.

Traveling farther afield will require a trip in a futuristic aircraft, which will be more like a bird than a plane. These shape-changing airplanes will adapt to flight conditions and provide passengers with the ultimate smooth ride. They will set records for quick journeys while using less fuel to do so, all thanks to smart materials. This is the future of the material world—and it's an exciting place to be.

The humble bicycle

Planned obsolescence—when manufacturers deliberately limit the lifetime of their products to encourage repeat purchase—together with impenetrably unfixable goods are contributing to an increasingly consumerist and throwaway culture. Although the bicycle, with its universal design and deliberately replaceable parts, gets close to a fully repairable product, even the best-loved machine will eventually end up with flaky paint, rusted parts, and collapsed tires. But in the future, smart materials will save our bicycles from the scrapheap.

Self-healing paint contains a resin-repairing agent encompassed in tiny microspheres that rupture when the surface is scratched. When this happens, the resin is released and fills the scratch, automatically repairing the damage. Self-healing tires could be made from vulcanized rubber that has been modified to have charged particles (ions) along the long molecular chain lengths. These oppositely charged parts of adjacent molecules attract each other to form strong bonds, which make the overall material strong and durable. If the molecules are pulled apart by a tear in the rubber, they can re-form the bonds spontaneously because of the simple fact that opposite charges attract. Today's puncture-proof tires involve a separate sticky sealant underneath the tread to fill punctures, but smart self-healing rubber is different because it involves a single component that can repeatedly seal itself.

Leave a bicycle out in the rain today and it won't be long before the exposed parts begin to rust. When this happens, the surface of the metal becomes more alkaline. Smart halochromic materials change color when the pH of their environment changes, like paintable litmus paper. A common example is phenolphthalein, which turns pink in alkaline conditions. A halochromic coating on anything from bicycle parts to bridges would allow for the early stages of corrosion to be effectively identified and treated before severe damage is done.

NASA has taken anti-corrosion coatings one step further, by developing a smart paint that not only identifies corrosion but contains microcapsules that release oily inhibitors in response to an increase in alkalinity to stop corrosion in its tracks. Autonomous prevention of rusting would be a big deal for national economies—astonishingly, the cost of corrosion in the US alone is conservatively estimated at around $500 billion a year, and might actually amount to more than twice that.

We may soon have the Curie brothers and their piezoelectric smart materials to thank for streetlights, road signs, and traffic lights, which will be powered by the roads themselves. The most common piezoelectric material used today is a manmade ceramic called lead zirconate titanate. This material produces an electric voltage when squashed because the atoms inside are arranged in an asymmetric crystal structure. To most people, the term crystal summons a picture of glistening transparent gemstones, but to a materials scientist, crystals are solids in which the atoms are packed together in orderly rows to make a three-dimensional repeating pattern. In fact, most gemstones are crystals, but metals, ceramics, ice, rocks, and some plastics are all also made up of crystals. The pattern of atoms that repeats over and over in the structure of most crystals is symmetrical—it looks the same back to front or upside down. Piezoelectric crystals have asymmetrical repeating units. Ordinarily, the charges on the atoms in a piezoelectric crystal cancel each other out; a negative charge here is negated by a positive charge nearby. But squeezing or stretching the asymmetrical repeating unit moves the atoms in such a way that their charges cannot cancel out. One side of the repeating unit therefore becomes more positively charged and the other more negatively charged. Multiply this squeezing and stretching effect over the millions of repeating units in the whole crystal and the result is a measurable electric voltage. Wiring up a piezoelectric crystal to a circuit allows for this voltage to be collected as useful electricity. These materials could be incorporated under asphalt to produce electricity when compressed by the weight of driving vehicles. This energy can then be stored in batteries and used to power roadside lighting. Several pilot studies of these systems have already proved promising, and the technology could even be incorporated into vehicle tires or the soles of shoes to produce electricity from motion.

In the future, cyclists, motorists, and local councilors alike will be relieved to find that potholes become a thing of the past thanks to self-healing concrete, a smart material that can detect flaws and repair itself autonomously. When concrete cracks, material on the inside gets exposed to atmospheric moisture or rain. Self-healing concrete contains embedded ingredients that are stimulated by water to fill in the crack. One example is a clay-based additive containing dormant bacteria and calcium lactate, a material familiar to anyone who has left old cheese in the fridge and then found it some time later with white crystals on the surface. The bacteria, activated by water, consume the calcium lactate and excrete limestone to fill in the crack and prevent further damage. This material could be used in roads, buildings, and other infrastructure, too, and would be particularly useful in those parts of the world prone to seismic activity.

In cold climates, cyclists need clothing that can adapt to body heat so that it is breathable on the bike but warm indoors. The solution may come in shape-memory polymers that change shape suddenly when heated. Polymers are materials with long, string-like molecules, such as rubber, plastics, and natural materials like proteins. The "remembered" form of a shape-memory polymer is set when it is initially manufactured. The material is then heated, moved into a temporary shape, and cooled. It remains in this shape until it is heated through the transition temperature, whereupon it springs back into its original position. These shapes are remembered by the material every time it is heated or cooled. Future cycling jackets could be filled with a fluffy shape-memory polymer liner that traps air like a sleeping bag in the cold but contracts in response to increased body heat to be less insulating during exercise.

Similarly, moisture-sensitive polymers change shape in the presence of water. When dry, the polymer is stiff, but water acts as a plasticizer, causing the material to relax. Such polymers can be used to make a fabric with miniature moisture-sensitive scales that stick out at right angles to the fibers when dry to allow breathability. When it rains, the scales relax to lie flat and overlap, providing an impermeable waterproofing layer.

We are familiar with the healing properties of our skin when damaged, a feature particularly useful in the event of an "unplanned" dismount from a bicycle, but in the future our ripped clothing will be able to do the same thing, thanks to self-healing textiles. These fabrics contain an unusual ingredient—a special protein that is found naturally in the "teeth" that encircle the suckers on squid tentacles and that can be made synthetically in the laboratory. This protein enables fabric on either side of a small tear to form new chemical bonds, allowing it to reseal itself in less than a minute with just the addition of water and pressure—a quick roadside pants repair. If only the ego was so easily healed.

A futuristic flying machine

Inspired by the flight of animals, Leonardo da Vinci imagined a human flying machine over five hundred years ago, centuries before anybody had even thought of the humble bicycle. His design involved jointed and flexible wings made from wood and silk, which were able to flap like a bird or a bat. Although our aircraft today resemble Leonardo's two-winged invention, they are rigid structures with just a few stiffly movable parts. Looking to the future, smart materials will release us from the inflexible constraints of current designs, taking us back to Leonardo's drawing board to make aircraft that are muscular, flexible, adaptive, and sensing.

Over the course of a flight, aircraft must be able to withstand many different forces, for which the traditional rigid shape is not always ideal. Planes of the future could have wings that can flatten or bulge to optimize aerofoil lift, and can fold, extend, twist or lie closer to the fuselage according to the stage of flight. These real-time adaptations will decrease drag and increase maneuverability, allowing for shorter take-off distances and optimized aerodynamics depending on live flight conditions. This would increase passenger comfort, shorten flight times, and reduce fuel consumption.

It will take a combination of all the "flavors" of smart materials to make such an aircraft a reality. The actuator components that physically move the wings will be made from shape-memory alloys like Nitinol, an alloy mixture of nickel and titanium, which moves between two preset shapes when heated and cooled. Lightweight shape-changing materials such as electroactive polymers that expand and contract when an electric voltage is supplied and removed will also feature heavily. Shape-memory polymers will be particularly important as the outer skin of the aircraft, since they can quickly switch from being strong and stiff enough to withstand aerodynamic forces to being extremely elastic and flexible to facilitate a morphing wing shape beneath.

Many of these smart materials will play a dual role as sensors, too; piezoelectrics and electroactive polymers generate measurable electrical signals in response to physical stress. The refractive index of optical fibers changes with temperature or when a force is applied, so embedding them in the aircraft structure not only produces a strong, stiff, and lightweight composite material, but one that can monitor damage, cracks, and dynamic strain during flight, too.

These smart composites could be used anywhere, from buildings made from smart concrete that alert engineers to potential failure points, to flexible electronics.

Futuristic flying machines could have human-level touch sensitivity thanks to quantum tunneling composites—smart materials that switch from an electrical insulator to a conductor when compressed. They comprise a soft rubbery matrix containing small particles of nickel. The matrix is an electrical insulator, and in the inactive state the conductive nickel particles are too far apart for the composite to conduct electricity. When it is activated by being compressed, the nickel particles move closer together and electrons "tunnel" through the insulator so that the overall material becomes electrically conducting. In quantum mechanics, the location of an electron is described in terms of the probability of it being in a particular place rather than in terms of its exact coordinates. When an electron in a nickel particle approaches the barrier of the insulator, there is a very tiny, but non-zero, probability that instead of bouncing off it will get through to the other side. Quantum mechanics says that with enough electrons trying to pass through and a non-zero probability of them doing so, some will get lucky and end up on the other side. These exotic smart materials have already been used by NASA robots to sense how tightly they are gripping an object, and could also be used as novel touch screens and, for amputees, prosthetic limbs that can actually feel.

Our smart material legacy

All the smart materials explored in this chapter have been proven to work on the laboratory bench. However, there are a number of problems that need to be overcome before we get to experience smart materials in our daily lives. In many cases, the response times are too slow and the materials are too delicate or unstable. Their performance can also diminish over time, and incorporating them into working devices is a difficult task. Often the switch threshold of the stimulus is difficult to control. Some of the materials are toxic and, as with many new technologies, prohibitive cost, upscale of manufacture, and availability of raw materials are holding them back from being widely used at present.

However, I'm optimistic that many of these problems will be overcome by continued research, and in the same way that the internet transformed how we interact with information, smart materials promise to overhaul the way we interact with the material world. By definition, an object is an unthinking thing that has actions done to it. Taken in isolation, even smart materials are just simple one-trick ponies, flip-flopping back and forth between two states given an on/off stimulus. But combine them to construct an aircraft that can generate and store energy, sense itself and its surroundings, self-assemble and self-heal, adapt to the current environment, and communicate with others of its kind, and suddenly objects begin to sound very much more alive.

Smart materials therefore invite us to consider some big questions. Are they even necessarily for the better? Is it right that we should prefer to increase the complexity of simple objects to make our own lives more convenient if this uses more precious energy and scarce resources to produce, use, and recycle them? But if they result in our using less energy or fewer resources, then that creates its own problems, too; self-healing smart materials will present a huge challenge to manufacturing and commerce by extending the lifetime of goods. How will our economies adapt? Will smart technology be available only to a wealthy elite? If we start to rely on smart materials to automate our lives, do we risk becoming dependent on them and losing our capacity to think critically or independently? In artifacts, materials tell the story of humanity. One day, museum pieces made from these now state-of-the-art smart materials will be all that is left for us to be remembered by. What will they say about us?

In my opinion, the positive impact of smart materials on the lives of people around the world far outweighs any potential challenges and negative implications. For example, the materials required for a futuristic airplane are the same as those that could make a brain-controlled moving, sensing, and self-healing prosthesis for an amputee. If adaptive textiles and self-healing infrastructure can mitigate the effects of an increasingly unpredictable climate for people in the worst-affected regions of the world, then they are technologies worth pursuing. The relationship we have with materials is personal, complex, and telling of the prevailing thoughts and ideas of a particular epoch. When we are gone, I hope that these smart materials will speak of a human race that is sensitive, adaptive, and resilient in the face of often challenging and changing environments, reflecting the materials themselves.

13

Energy

Jeff Hardy

I have a confession to make: I'm an energy geek. I get excited about energy—from the cutting-edge science, like harvesting energy from our own motion, through to the most mundane, like the humble gas boiler. That makes me unusual. For many, our relationship with energy is rather indirect—we treat it as an essential service, something that should always be there when we flick a switch. This is not the case in the developing world, where two billion people aspire to have better access to energy. In the future, our relationship with energy is going to change—our demand for energy conflicts with the consequences of climate change and other constraints. If there's one thing that's certain, it's that energy—and particularly finding a source of unlimited clean energy—will be one of the most important issues facing humanity in the future. Rather than fear uncertainty, we should embrace it; this chapter explains why we should all be excited about the future of energy.

Energy and greenhouse gases

You will have read Julia Slingo's chapter about climate change (Chapter 3), so you know how vital it is that we reduce greenhouse gas emissions globally. According to the International Energy Agency, energy production and use accounts for two thirds of the world's greenhouse gas (GHG) emissions. Given this contribution to the problem, it has to be a massive part of the solution.

How do you decarbonize energy? The answer is deceptively simple: Stop burning fossil fuels. In practice, of course, that's much more difficult to achieve, as fossil fuels are synonymous with modern life. They literally fuel our lives. It wouldn't be fair, or popular, to tell people to put their daily lives on hold while we come up with an alternative. Nor would it be fair to halt development in those countries where access to energy is limited, even where it is driven by fossil fuel use. Therefore, we need a plan to reduce our reliance on these fossil fuels in an orderly fashion.

Firstly, we need to be more *efficient*. The cheapest way to reduce GHGs is to not burn the fossil fuels in the first place. In the US (and elsewhere), we waste a lot of energy. For example, two thirds of the energy used to provide our electricity is lost during generation and transport through our networks. Similarly, heat is lost from poorly insulated buildings. So, if saving energy is both easy and a good thing, why aren't we better at doing it? The answer is partly because people just aren't that interested in energy efficiency, and bribery (e.g., grants/subsidies and the like) doesn't work for the majority. There's another problem as well: Being more efficient saves people money, causing a "rebound effect" whereby they tend to spend the money saved through energy efficiency on something else that uses energy and generates GHGs. For example, you can afford to heat your home to a warmer temperature, thus burning more gas, or you spend the money on a lovely vacation, flying by jet.

The second solution is to be *smarter* in how we run our energy system. Most energy systems are run on the basis that the supply of energy follows demand for energy. This is particularly true for electricity, where supply and demand must be kept constantly in balance. For electricity systems, there is a merit order of technologies, which includes those that are always on (called baseload, such as nuclear and coal power), variable (such as solar and wind power), flexible (such as gas-fired power stations), and peakers (such as diesel generators, which can turn on at a moment's notice, but are expensive). If you want clean clothes, you put the washing machine on. To meet this demand, someone in the energy system turns on or turns up a power station. If this is at a busy time, for example when everyone is washing clothes, then that power station is likely to be powered by fossil fuels (a gas power station or diesel generator). A smart energy system is where demand follows available supply. In the above example, our washing machine might ask: "When do you want your clean clothes by?" The washing machine will start when energy supply is available (as long it meets your deadline!), meaning that no additional power station needs to turn on.

The third is to *substitute fossil fuels* with other technologies or fuels that have lower or zero GHGs.

For electricity, you can substitute fossil fuel power with low-carbon alternatives such as renewables, wind, solar, and biomass (burning trees or other stuff that grows), and nuclear power.

An alternative is to use carbon capture and storage (CCS). This involves attaching a large chemical plant to a power station that captures most of the carbon dioxide from the exhaust gases. This carbon dioxide is then transported by pipe and injected into something like a disused gas or oil field, where it is trapped forever (in theory). If, instead of burning fossil fuels in a CCS power station, you burn biomass and capture the carbon dioxide, then you can have negative carbon dioxide emissions! This is because when a tree grows, it absorbs carbon dioxide out of the atmosphere—so if you burn it and capture the carbon dioxide, you reduce the carbon dioxide in the atmosphere (as long as you are replacing the trees). Not all biomass is carbon neutral, though. It depends on how the biomass is grown, harvested, replenished, and transported, and the land that is used.

For heat, there are various options. Today we burn lots of natural gas (methane) to keep us warm. We could substitute this natural gas for another that doesn't contribute as much to global warming. For example, we could make biogas from biomass, which should have lower net GHG emissions with the caveats discussed above. Another option is to use hydrogen gas. When hydrogen burns, the by-product is water, so it's very clean. The problem is that most hydrogen is made from natural gas in a process called steam methane reformation, which produces carbon dioxide. We can instead make it by splitting water into oxygen and hydrogen with electricity, but that's still rather pricey, since it requires expensive catalysts like platinum and palladium, and the electricity must be paid for.

For transportation, fossil fuels are king (for now—one day we may be able to use hydrogen), although it is becoming normal to see electric and hybrid electric vehicles on the road, and they are becoming increasingly popular; there are over a million in the world now. A challenge is our expectation that a car should be able to travel hundreds of miles on a single tank and "refuel" in minutes. Batteries and charging stations are improving rapidly, but there is still some way to go.

Perhaps more so with transportation than electricity and heat, behavior is important. Today American drivers are in their cars about 3 percent of their lives. The promise of on-demand transportation, like autonomous vehicles, smart public transportation, and better cycling cities, as well as virtual-meeting technologies could, in the relatively near future, change our whole relationship with getting from A to B.

Disruption eruption

A disruptive technology is one that displaces an established technology and shakes up the market, or a groundbreaking product that creates a completely new industry. In energy, disruption is already happening.

For a start, things are becoming more efficient. Across Europe, energy labels and product standards have saved the equivalent of the entire annual energy consumption of Italy. As a specific example, the humble incandescent light bulb is being replaced by Light Emitting Diode (LED) bulbs, which use around one tenth of the energy. Smart technologies are helping us become more energy efficient, too, like smart thermostats that are learning our energy use habits and optimizing our home heating to save money (though, of course, there is always that "rebound effect" mentioned earlier).

Energy technologies are also becoming more affordable for us to own. Solar photovoltaic (PV) panels are getting cheaper as global deployment gathers momentum. Solar PV has a "learning factor" of around 21 percent, which means that for each cumulative doubling of the product's sales or installations, the price falls by about 21 percent. There are now something like 300 gigawatts of PV globally. Just ten years ago, this figure was closer to 10 gigawatts. (To explain, if I turned on everything in my house, I would need about 10 kilowatts of power, and since a gigawatt is one million kilowatts, this means that one gigawatt could handle one hundred thousand houses at peak time.) This is creating millions of "prosumers," consumers who also produce electricity. The next big thing is expected to be batteries to store electricity in our homes.

Data, data, everywhere

It's becoming ever clearer that we live in an era of "big data." Everything is collecting and transmitting data: our phones, smart meters, sensors, and the growing Internet of Things. Big data has already been used to improve marketing of products, predict earthquakes, and nudge patients who are likely to forget to take medication.

Data (big or otherwise) combined with machine learning (basically, algorithms that can learn from and make predictions on data) has applications for energy. It has also proven to reduce energy demand by optimizing processes, like refrigeration in supermarkets or cooling in data centers. The potential is much greater, though, as it could be a key component of using energy in a smarter way by optimizing supply and demand constantly in real time.

A final piece of the disruptive puzzle is the emergence of new and secure ways to process peer-to-peer transactions. Blockchain technology, which underpins the Bitcoin virtual currency, is a secure, distributed ledger that records transactions between parties. Using blockchain, or an equivalent, creates the opportunity for two or more parties to conduct a secure financial transaction without the need for intermediaries like banks. This is important for energy, as it could enable secure peer-to-peer energy transactions—for example, I could sell the electricity from my PV panels directly to my neighbor with no middleman involved.

Generation connect

It's not just technology that is changing; we are, too. Our relationship with energy could well change in line with our future values, attitudes, and behaviors. We can already see differences between generations.

Generation Y, called the millennials, born somewhere between the late 1970s and the 1990s, is an internet-connected, technology-savvy, liberal-minded generation. Surveys indicate that millennials live their lives through mobile technology and are more interested than previous generations in distributed energy technologies such as solar photovoltaics. As such, they are likely to embrace the disruptive technologies coming to energy.

Generation Z, born in the age of the internet, emerged, device in hand, into a social media and mobile world. They have grown up in tough times globally, including financial crises and the apparent increase in security threats. Surveys find them to be pragmatic, industrious, and socially and environmentally responsible. They adapt to innovation and disruptions and expect instant responses and services from the world.

Drivers of change are important, as they tell us something about how our relationship with energy could evolve in the future. We've already explored some trends that could make our relationship with energy more personal. Let's breathe some life into this through three vignettes of the future that could come to pass during the lifetimes of generations Y and Z.

Smart life

Consider this future scenario. Your time is precious—so precious, in fact, that you've outsourced responsibility for running your household to a company called "Life-back." In return for a monthly lifestyle bill, the company keeps you warm, fed, entertained, mobile, and productive. Your home is smart—you can speak to it and it can talk back—which is helpful when it comes to ordering groceries and the like. It's full of monitors and sensors that constantly optimize your living environment. Your home is well insulated and has photovoltaic panels on the roof and a battery in a cupboard. Occasionally, when you need heating, a small heat pump does the job. You didn't pay for these; the company did, and you are paying them back over time from the money you save on your energy bills. You don't own a car. Your smart home often knows when you need to be mobile and ensures the optimal transportation mode is available. Sometimes this is an autonomous electric vehicle arriving at your door. Other times it's a reminder when the next public transportation is arriving or a prompt to put your cycling gear on. When you signed a contract with the company you agreed that they could take actions on your behalf. This includes switching between who supplies the small amount of electricity you need as well as managing electricity demand in your home to help out the electricity networks when they are constrained in some way. You are a very active participant in the energy market, albeit with someone doing it on your behalf.

You are now also an energy trader. Welcome to the world of peer-to-peer energy! The old model, where you got your energy from a supplier, is a thing of the past. Today you cut out the middlemen, dealing straight with other generators and consumers. So how does this work? Your home is a power station. You have solar panels on your roof, perhaps a wind turbine in your garden, and a battery to store electricity. If you need one, your car is almost certainly an electric vehicle. Your peer-to-peer market allows you to buy or sell things, in this case electricity. Transactions are recorded and paid via secure public ledgers based on blockchain or similar technologies. Through this, you can buy or sell electricity securely. Who you choose to trade with and how you choose to trade is up to you. You might want to maximize your profits, selling energy when the price is higher and buying when cheap, or maximize your consumption of green energy. You can be totally hands-on, or leave it up to a third party to optimize your strategy on your behalf. It makes sense to insulate your home and install smart devices that control home energy demand. Together with batteries, this means you can vary your home energy demand to optimize your trading strategy. It also helps you prioritize things like charging your electric car or heating your home. You can be as active as you like in the energy market—the rewards scale with your activity.

You've decided to take back control of energy in the most direct way by disconnecting from the grid. To be self-sufficient you've had to consider electricity, heat, and mobility. You've minimized your demands by making your home ultra-energy-efficient. All your home devices work on DC rather than AC current, which means they can use the power from your solar panels directly. For additional heating and power, you've bought a fuel cell system that converts household waste (including sewage) into electricity and heat. For communication, connectivity, and entertainment, you have mobile internet and connected devices. While you are out and about, you harvest energy from the environment through tiny solar panels integrated into your clothes and through motion-harvesting devices in your shoes. This allows you to charge a battery you carry with you, meaning your mobile devices are never wanting for power. Occasionally, you might find yourself short of power, but that's something you've come to live with. You could always ask your neighbor if you could borrow some. . . .

Sometimes individuals may get together as communities to take control of aspects of their energy. These can be so-called communities of interest or place (or both where these coincide). A community of interest might be where people come together to collectively get a better deal for energy supply (so-called "collective switching"), or perhaps crowdfund an energy project. Communities of place are local communities, like a village or town, where residents get involved in buying and operating local energy infrastructure, like solar PV panels, and sharing the power and profits locally. In some cases, communities have purchased the whole local energy system, including the electricity network and heat network.

But what if you aren't already on the electricity grid? This is the case for one in six people on our planet today. What if you could leapfrog the need to build a countrywide electricity grid, in the same way that mobile phones negate the need for landlines? This is already a reality courtesy of what we call microgrids, which involve a combination of solar PV (or other electricity generation technologies), batteries, control systems, and wires to connect a community together. In other words, it's all you require to provide a basic energy service to a community off the grid. It provides light at night and allows communities to charge devices such as mobile phones. Basic services like these have huge social and economic benefits. In fact, improvements and cost reductions in solar and battery technology could negate the need for a national grid to ever develop. If you combine microgrids with mobile networks and mobile internet, you are leapfrogging three networks that have taken decades to develop in other countries. Other developments, such as cashless payment systems, where citizens can pay for goods using just a mobile phone and a personal identification number linked to biometric data, such as a fingerprint, could completely replace hard cash. This enables newly connected communities to do things such as take out business loans or pay for goods and services, creating new economic opportunities. Perhaps in countries where there is no national grid, these advances can mean that such a thing will never need to be built.

Thinking global

Let's zoom out from individuals and communities and think in terms of intercontinental or even global energy. Of course, energy is already global in that we move vast amounts of resources between countries. But could we do the same with electricity?

In many ways, we already do. For example, the US and Canada share interconnected power grids and gas pipelines. And most European countries are connected—even the UK is linked to France, the Netherlands, and Ireland by interconnectors (undersea electricity cables and pipes carrying natural gas). But could we do that on an even grander scale?

DESERTEC is a rather sunny supergrid proposition. The idea behind it is to generate power at wind and solar PV farms and concentrated solar power plants (CSP) in and around the Sahara desert, then send the excess electricity across to Europe through a new high voltage direct current (HVDC) grid via Spain and Italy. These CSPs deserve a quick explanation. Imagine a field of mirrors pointing at a collector on a tower, a little like focusing sun through a magnifying glass. In this case, the concentrated beam of solar energy generates temperatures of up to 1,832 °F, which drives a steam turbine just like a traditional thermal power plant. Any excess heat from CSPs can be stored by melting salt in giant tanks. This heat can then be used to generate electricity at night.

Scientists have also imagined putting solar panels in space and beaming the power back down to Earth. This sounds like the plot of a James Bond film, but it has been around as a concept since the 1970s. It's called space-based solar power (SBSP), and there is sound logic behind it. Around 60 percent of solar energy is lost on its way through the Earth's atmosphere, so putting solar panels outside the atmosphere increases available solar energy massively. Once you overcome the question of how to get solar panels into space, the next challenge is how to get the energy back to Earth. Microwave or laser beams are the best bets for this. Finally, you need something to point these beams at—a "rectenna"—perhaps several miles wide, which receives the beams and converts them to electrical power. So, other than the issues of pointing high-energy beams at the Earth, there's the small matter of sorting out the political and economic dimensions of who owns and pays for the energy. . . .

One for the road . . .

That's quite a canter through the future of energy. The details of all these predictions are almost certainly wrong, but I hope that you can understand why our relationship with energy is likely to change in the future, one way or another.

Let me leave you with a thought experiment.

Suspend your disbelief for a moment. What if energy were entirely free? What if there were no wires and pipes? What if you could pluck energy from the air like we do wirelessly with data? This sounds rather farfetched, but it could be possible. Free (or very cheap) energy could come in the form of super-cheap solar power (perhaps beamed down from space) or other renewables, or perhaps we will finally crack nuclear fusion (this is where we mimic how the sun works in a power station on Earth). Wireless transmission is also more common than you may think. Electric toothbrushes and some mobile phones are charged by inductive charging, so perhaps that technology or something similar could be used. To move power around over large distances, we could always use the laser or microwave beams from the space solar power example. My point, however, is not about technical or economic feasibility—it is about how this would change the way we interact with energy in the future. If energy were free, like the air we breathe, how would that change our relationship with it?

14

Transportation

John Miles

Sitting alongside such challenging subjects as Artificial Intelligence, demography, genetic engineering, and transhumanism, transportation, with its mundane images of buses, trains, and overcrowded airports, seems a bit out of place in a book that asks such important questions as "What's next for humankind?" But, historically, transportation has had a profound influence on the way we live and, with a little imagination, we might begin to see all sorts of ways in which it could continue to surprise us and shape our way of life in the future.

The basic needs of humankind are generally accepted to be a relatively short list that includes food, water, shelter, protection from environmental threats, and love. This list does not include transportation. But despite its absence from the classic list of human *needs*, transportation has been a crucial *enabler* for many of the critical stages in the development of our modern civilization. The invention of the wheel, leading to the introduction of the animal-drawn cart and thence to all forms of motorized transportation, is the simplest and greatest example. But, equally, the building of roads, canals, and railways has enabled society to develop in a manner inconceivable prior to the arrival of each of these concepts. To catch a glimpse of the next big steps in transportation therefore requires us first to understand how and why the history of road, rail, and air transportation has had such a profound influence on the way we live today.

Cost, time, capacity, and convenience

New transportation developments come in many different forms. Some are spectacular (Concorde, for example, and the Apollo Moon missions) and some, in retrospect, are very "ordinary" (the canal boat and the Model-T Ford). These examples suggest that the impact that new forms of transportation can have on the development of society is not related to the complexity or cleverness of the technology introduced, but is a reflection of some more basic property.

Let us consider three basic attributes that characterize all successful transportation systems. The first is affordability (or cost). Nothing discretionary permeates to the grassroots of society unless it is affordable and, in this respect, affordability is the clearest requirement for a successful transportation system. The other two key attributes are journey time (speed) and transfer rate (capacity).

Perhaps the greatest historic driver for transportation innovation has been the military. For the military, cost is inconsequential provided that the advantages of speed and capacity are assured. So it became commonplace for the Romans to build roads that enabled large numbers of troops and equipment to move swiftly from one land base to another regardless of the cost of construction. In doing this, the Empire created connections between townships that would subsequently become the vascular system for economic development—first at regional scale, and then at national and continental scale. A similar story applied to shipping. In both cases the initial reason for the development was the military advantage conferred, but the subsequent (and arguably, much greater) advantage was a general easing in the movement of people and goods. Over time, this led to increasing trade over longer distances and in turn to wider economic development—regional, national, and international. In the end, we can see a clear link between the development of road- and sea-based transportation systems and the establishment of nations, empires, and trading blocs. But the transition from military requirement to economic enabler is not automatic. We only have to look at the history of supersonic flight to see that, in some cases, a stellar military performer turns out to be more or less irrelevant to the subsequent pattern of socioeconomic development. Concorde has not become the universal travel solution its originators hoped it would. Why not?

In my view, the answer is that it didn't deliver a workable combination of our three transportation attributes—cost, time, and capacity. The journey time advantage (speed) was extraordinarily high, but this advantage was not sufficient to justify the huge cost per passenger mile, nor to overcome the very low passenger transfer rate (with a maximum of only 128 passengers per flight). In reality, the aviation winner was the jumbo jet, where increasing the number of passengers on each aircraft simultaneously reduced the cost per passenger mile and increased the capacity of the overall system (despite the slower speed).

The jet aircraft has been very effective at transferring people over long distances, but it has been less effective at transferring goods. Moving goods across the world is much better achieved by sea transportation, which scores extremely well on cost per unit of advantage (in this case, cost per ton-mile shipped) and capacity (tons per day or year), but very poorly in terms of journey time. Prior to the introduction of aircraft, there was no other choice for intercontinental travel, so the ocean-going vessel had limited success as a means of passenger transportation. But the journey time was so long that, in all but exceptional circumstances, the journeys had to be one-way only. This enabled emigrations to take place from Europe to the United States, Canada, South Africa, Australia, etc., and it also allowed movements in the opposite direction from the former colonies to Europe. Sea travel therefore had an enormous impact on the development of our modern society. Today, however, it has become a thing of the past. The ship would never have opened up the universal two-way freedom of movement conferred by the airplane. However, for non-perishable goods, journey time is of little significance. As a result, sea-borne transportation remains very important to the shipment of freight. The introduction of containerization and the building of ever larger ships have led to an enormous increase in the intercontinental movement of goods, and this has become a massive pillar of globalization.

While cost, time, and capacity might have been the convenient yardsticks against which historical transportation developments could be judged, they may not be the right yardsticks to continue using in the future. In the past, we might have chosen to take the train for a commuter journey to work on the basis that it was quicker than a car journey—and maybe we could do some of our work as we travelled. But if, in the future, we can work from the car because we are connected by mobile phones and internet and the car drives itself, we might change our minds. In other words, the emergence of mobile communications, on-board computing, and autonomous control systems might combine to make the choice of travel mode more a function of *convenience* than *time*. In my view, the way that technologies might come together to transform the way we think about travel is a revolution in the history of transportation and further underlines the hidden power of the digital revolution.

So what are the factors that might characterize the future transportation possibilities for us and our children?

Environmental impact

Currently, the average annual carbon footprint for a passenger vehicle in the US is around 4.7 metric tons of carbon dioxide. While the current administration is sending mixed messages about its goals to cut emissions, in many countries there is a legal requirement now to reduce the national carbon footprint to 20 percent of the 1990 level by 2050. Given the likelihood that some forms of energy consumption may never be able to reach this level (air transportation, for example), road transportation may well have to do better. For road transportation, could we therefore reach 10 percent of 1990 levels, or even zero CO2 emissions, by 2050?

Most commentators agree that, in the longer term, the days of the internal combustion engine are numbered. A window of opportunity has therefore opened up for those with imagination. Cue the arrival of the Tesla motor company, an enterprise started from scratch in Silicon Valley in 2003 by a team of entrepreneurs and engineers with a strong vision for clean but exciting personal transportation. After a slow start, and the later arrival of the iconic entrepreneur Elon Musk, the company has since produced an attractive family of battery-powered electric cars and is now well on its way to selling one hundred thousand vehicles per year. The stock value of Tesla has now surpassed that of Ford—an event that would have been inconceivable not too long ago. This is a dramatic illustration of the new disrupters at work—companies with no history in the motor industry that refuse to play by the established rules. This could not have happened at any other time in the seventy-year period between the consolidation of the early automotive pioneer companies and the present day, because even the most dedicated pretenders would never have had deep enough pockets to survive the industry's barriers to entry. But the self-styled "tech-companies" are different; within Silicon Valley they are capable of raising huge resources through shareholder offerings. Thus the rise of Tesla brings with it the sudden possibility of a sea change in the industry. Whether the disrupters will ultimately go bust or will succeed in their showdown with the traditional motor industry, they have already had a lasting impact. Over the past several years, almost all of the established giants have declared their intention to produce a range of all-electric cars, and few doubt that they will deliver on that promise before long.

At the moment, the convenience and capacity attributes of the electric car are hardly any different from those of the conventional fossil-fuel combustion-engine-powered car, while (for the moment, at least) the affordability is worse. So while this change is more likely to result in a gradual like-for-like swap of battery-powered cars for conventional ones as the prices of batteries fall, it is not likely to change our socioeconomic landscape in any dramatic way. But some other aspects of the disrupters' arrival might have a much bigger socioeconomic effect, which I will return to later.

Public transportation and intelligent mobility

The digital revolution has opened up people's perception of public transportation, and the concept of "intelligent mobility" has emerged as an accepted wisdom within the world of transportation professionals. This means that making a journey only by car, or bus, or train, is replaced by the idea that a traveler might get from A to B using a combination of whatever public transportation options are available. The resulting so-called "multimodal journey" has always been the transportation planners' holy grail, but it has been frustrated in the past because the idea of standing for half an hour in the rain at a bus stop, or walking half a mile to the nearest underground station mid-journey, is too horrible to countenance for the average traveler. For too many commuters around the world, public transportation has come to be seen as the lowest form of traveling, something to be shunned unless there is no other choice, and the car has therefore become the default choice for those who can afford it, despite the traffic jams, parking difficulties, and other inconveniences.

In the brave new world of intelligent mobility, public transportation could be seen in a new light. The dream is that traveler information systems connected to flexible, on-demand public transportation systems, from trains to buses to bicycles, could mean that intra-journey transfers can be made without any delays and uncertainties. Public transportation could suddenly become capable of providing a quick and reliable door-to-door service without the inconvenience of having to drive under stressful circumstances on congested roads or struggle to find a parking space once you have arrived.

Autonomous vehicles

The increasingly rapid development of autonomous vehicles is another area in which the disrupters are threatening the established order of the automotive industry. The appearance of Google, Apple, and Uber on the scene is creating a great deal of tension. Once again, whether or not these companies ultimately displace the established players is not really the issue; they have already created a huge change in the direction of the industry, and they will leave their imprint on the future because of it. The winner of the race to deliver the first fully autonomous vehicle will have won an academic accolade, but the important point is that the race has been started and both the establishment and the disrupters are now moving at an accelerating pace.

This technology has a clear potential to become a game-changer for two reasons. In many developed and developing cities, from New York to Mumbai, there is acute congestion and an increasing pressure on politicians and on public funds to provide more road and rail capacity. Building new infrastructure is expensive and inconvenient, and the timescale for planning and implementing it can be measured in decades. A few simple calculations, however, suggest that autonomous vehicles could hugely improve the situation through more disciplined lane control, tighter vehicle spacing, and fewer minor accidents because of driver inattention. The extent of these benefits could easily outweigh the increased numbers of journeys that the use of shared autonomous systems would undoubtedly generate. Achieving a reduction in congestion levels is the same as providing increased capacity, and the promise of a significant increase in capacity with very little capital outlay and near-zero inconvenience is a goal to be pursued with some vigor.

The second reason why transportation watchers should take autonomous vehicles seriously is the promise of "time recovery." This bit of jargon refers to the fact that, once the vehicle drives itself, the (former) driver will be free to use the journey time for other purposes. This might include office work, leisure pursuits, or simply relaxing/sleeping. Each of these choices has the effect of removing the unpleasantness of a slow journey. If you can do something that you consider to be valuable with your time en route, there is no reason to become stressed if the journey is slower than intended, particularly if the information systems at your disposal are able to warn you of the delay in advance and keep you up to date as you progress toward your destination.

So where might the real stars of the future emerge? The honest answer is, I don't know. But there are some very interesting candidates floating around, and I can make some informed guesses.

First, the dominance of the car for relatively short journeys in the urban and suburban environment will be weakened (but not quite eliminated). This erosion of the car's supremacy will come about because of the vastly improved levels of public transportation service that can be provided as intelligent mobility solutions begin to mature. The *convenience* of using public transportation systems will dramatically increase, the capacities will grow in response to demand, and the cost per passenger journey will plummet as a direct result of this (just as we saw happen in air travel with the jumbo jet). The critical element for success in this scenario, of course, is that public transportation will provide a more attractive means of traveling from A to B than the private car. Attracting passengers by offering a better service is a much more powerful driver than threatening them with penalties if they don't stop using their favorite form of transportation (currently, the car). A more attractive public transportation service will be achieved by ensuring reliable, seamless, multimodal journeys in which the remaining inconveniences of making intermodal changes are outweighed by the advantages of avoiding stressful driving experiences like traffic jams and destination parking.

Second, the dominance of the car for medium-distance journeys (a couple miles to a couple hundred) will be reinforced rather than diminished. This will come about as a result of autonomous vehicles coming to market, which means we will enjoy significantly increased transportation capacities and see a radical reduction in road accidents, both of which will be achieved without putting a significant additional burden on the public purse. This will not so much "transform" our current socioeconomic environment as simply ease our current difficulties, but it will be an important development nevertheless.

My third prediction concerns the much-vaunted prospect of near-sonic land-based transportation ("Hyperloop"). This idea has received a lot of press coverage since it was originally proposed by Elon Musk in 2013. It is, in effect, an electric airplane in a tube. Imagine the effect that a fast and frequent service between distant cities could have if the frequencies and journey times were comparable to those of a modern urban metro system. Imagine being able to travel between New York and Philadelphia in the same time that it takes to travel between the Brooklyn Bridge and Times Square on New York's subway; similarly, Musk has mentioned travelling from San Francisco to Los Angeles in forty-five minutes. This would have the same economic effect at national scale that the London Underground or Paris Metro have on those capital cities—namely, that the whole area served begins to function as a single, homogeneous economy. Similar arguments may be made for other small to medium-sized countries. But—like Concorde—it will only work if we can achieve low journey costs and high passenger transfer rates at the same time as delivering very short journey times. If not, it may only be a niche solution for travelers with expensive tastes—a model that has little chance of changing the socioeconomic order of things.

Beyond these thoughts, I can see little on the horizon that could make a transformative change to the way we live. No doubt low-emission (even zero-emission) road vehicles will come along, but they won't, in themselves, change our patterns of life. There will be developments in marine technologies, but while the cost and capacity parameters might continue to improve, the fundamental passenger transportation handicap of transit time is unlikely to be overcome. So this form of passenger transportation will continue to be eclipsed by the airplane for as long as airplanes are with us. And the opposite arguments apply to air transportation. There will be continuing incremental improvements to conventional aircraft but, in the absence of a dramatic change in capacities and flight times (without an increase in passenger journey costs), a lasting change will not occur. And the potential for achieving a transformative change in transporting freight by air will hang on the ability of airships to match the capacity levels of very large sea-going container ships—a very unlikely prospect.

The final transportation frontier, of course, is space (see Chapter 16). But now that you are familiar with the principles of the cost/convenience/capacity argument, I will leave it with you to work out just what this combination will look like, for whatever the answer, it will have a lasting effect on the way we live. . . .

15

Robotics

Noel Sharkey

There are many possible futures for our relationship with robots. Some are greatly beneficial to humanity, while others could lead to a dystopia. There is no way to tell which of these futures will happen. This is a field replete with crystal ball gazing and science fiction tropes. Amid the distractions, while governments are greedily eyeing the potential boost of billions of dollars to their economies, there are pressing issues that need to be dealt with now before Pandora opens for business.

We have been loitering on the cusp of a robotics revolution for a couple of years now. Since the 1950s, robots have been working in our factories painting cars, assembling components, and doing many of the dull and repetitive jobs faster than humans. But now these robots are greatly outnumbered by the newer "service robots." These are robots that operate outside the factory to work on everything from healthcare to the care of children and the elderly; from cooking and preparing food to making and serving cocktails; from domestic cleaning to agriculture and farming; from policing, security, and killing in armed conflict to monitoring and repairing climate change damage; and from robot surgery to robot intimacy and protecting endangered species.

These are exciting times for the industry. The lure of massive new international markets is driving governments and corporations to view robotics as a powerful economic driver, and funding is starting to pour into developments. Many companies and start-ups are creating a multitude of new robot applications in what is becoming a highly competitive market that will drive innovation.

Much of what happens next will depend on international regulation and on whether the engineers and the companies that develop and produce the robots take a more socially responsible approach. Pushing many, but not all, of the developments requires consumer trust. Without it, progress will be very much slower and investors will lose out.

Self-driving technology

One hot day in the Mojave Desert in Nevada in 2005, a Volkswagen Touareg car without a driver shimmered into the history books, scooping up a $2 million prize. It was the first vehicle to cross the finish line in the DARPA grand challenge, a 132-mile desert road race that took the car, named Stanley, 6 hours 53 minutes 08 seconds to complete. Another five autonomous vehicles rolled over the finish line within minutes of each other.

This was an extraordinary step forward in robotics. In the previous DARPA challenge a year earlier, the best distance any of the fifteen competitors could achieve was 7.1 miles, 124.9 miles short of the finish. The winning team was quickly snapped up by Google, which began its famous Google Self-Driving Car (SDC) developments.

Now, many companies are getting in on the act to compete in the new transportation market. There are already self-driving trucks carrying out daily deliveries of materials from mines in Northern Australia. There are self-driving tractors being used for more efficient food production and autonomous buses being tested in the US, Denmark, and Japan. Uber has also added self-driving cars to its fleet of taxis in the US. Of course, for the foreseeable future, all such self-driving vehicles will still have a human occupant who can intervene if necessary.

A major push for self-driving is energized by the belief that autonomous cars will dramatically reduce the number of road deaths. The argument goes that they use sensors to detect other cars and pedestrians more quickly and reliably than a human driver and so are better able to maneuver to avoid them. In addition, a robot car won't get sleepy, can't get drunk, and won't be distracted by a screaming child in the back seat.

But not everyone is convinced. Speaking after Nevada changed its laws to allow self-driving cars, Trooper Chuck Allen from the Nevada Highway Patrol said, "When you factor in the safety of others by having mechanical devices dictate your speed and thoroughfare, it does cause some concern." Although accidents with autonomous vehicles have been rare, there have been some serious setbacks and a couple of fatal crashes. "Drivers" are required to be alert and attentive so that they can take over in the event of unforeseen situations—for example, if there are temporary road signals, hazards that can only be interpreted properly by humans, or, say, a police officer signaling to drivers to slow down because of some incident farther up the road. But spending hours doing nothing other than observing can lead to an automation bias where the driver comes to trust and rely upon the car, and is thus unprepared for hazards when they do arise.

Then there are ethical questions, such as what a driverless car should do in the event of having to make a choice. Should it crash into the old man on a bicycle rather than the pregnant woman with two children in the other car, or avoid them all by swerving into a wall and killing its own occupant? Although these questions are interesting, they are far removed from the current sensing abilities of the car. Apart from detection accuracy, the car would have to "know" the road surface and whether something has been spilled on it. It would need to do the impossible in sensing what is hidden from the sensors.

It will still take some years to iron out all of the difficulties, but we can expect our highways to have changed dramatically by 2050. We might reduce accidents massively if we took humans out of the loop altogether and the cars communicated directly with one another and with sensors on the road to alert each other about accidents ahead and automatically respond to prevailing conditions. Of course, the one downside is that they may all still be hackable.

Domesticating your robots

Even if they tend to be on the duller side of the technology, it is true that many of us already own a robot. Domestic robots are not the humanoid servants from science fiction. They don't look like C-3PO. Some of them look like fat Frisbees on wheels and carry out the tedious tasks that few want to do, such as vacuuming, window cleaning, mopping, and clearing the gutters and tackling sewers.

According to the International Federation of Robotics (IFR), there were 4.7 million robots sold for personal and domestic use in 2014. In 2015 that figure increased to 5.4 million with sales worth $2.2 billion. And this massive global market is unlikely to shrink anytime soon: the IFR forecast forty-two million units sold for 2016–2019 (at a conservative estimate).

In the household of the future we will see more and more of our domestic duties like cleaning and washing being done by robots. There are also new developments, such as robots ironing and folding towels and robots able to pick up laundry, put it in the washing machine, and then unload it again. These robots will not be ready for production for a while, but we can expect them in our homes within the next twenty to thirty years, if not before.

There have also been rapid strides in robots developed for professional cooking, such as burger, pizza, and sushi makers. And of course there are also robots that can make cocktails incredibly quickly. It doesn't take much imagination to predict that these will come down in price rapidly within the next twenty years, so that they can be available domestically.

I called this the "duller" side of robotics because it takes on the dull, everyday, tedious, repetitive tasks that fill much of our domestic lives. But this will not be so dull for us if it means we can put our feet up and have robots working as part of our connected "Internet of Things" homes (see Chapter 8). The big problem though is that, unless you live in a palace, there is likely to be no room for all of these appliances in your home. The really big development that we need in order to get all of this working is the multipurpose robot. However, there is no sign of that yet, and in any case the smart home of the future may well have the robots incorporated into our appliances and built into the structure of our dwellings. For example, an oven may detect that you have left the room twenty minutes earlier and turn off the stove beneath your frying pan.

Robots that care

The number of robots used for assisting in care has increased rapidly in the past decade. In 2014 there was a massive six-fold increase on the previous year in assistive robots for older people and people with disabilities, according to the World Federation of Robotics. This is something that could have a very positive impact if we get it right—and a serious negative impact if we get it wrong.

What we need to be vigilant about is the development of robot "nannies." A number of manufacturers in Asia have been attempting to fulfill this dream. These robots are equipped with video games, quizzes, face and speech recognition, together with limited conversation to capture pre-school children's interest and attention. Visual and auditory monitoring for caregivers is designed to keep the child from harm. These robots are very tempting for busy professional parents and their prices are falling. Some parents are even using cheaper ones, such as a Hello Kitty robot, which costs nearly $3,000.

An ageing population has created an even greater drive for the development of robots to help care for fragile older people. Japan is already en route to deliver robot-assisted care with examples such as the Secom "My Spoon" automatic feeding robot, the Sanyo electric bathtub robot that automatically washes and rinses, Mitsubishi's Wakamura robot for monitoring, delivering messages, and reminding about medicine, and RIKEN's RI-MAN that can carry humans and follow simple voice commands. Europe and the United States are facing similar ageing population problems and are now following Japan's lead.

As with any rapidly emerging technology, likely risks and ethical problems need to be considered. Many of the robot applications for children and older people could be considerably beneficial. For the elderly, assistive care with robot technology has the potential to allow greater independence for those with dementia or other ageing brain symptoms. This could enable the vulnerable older people to stay out of institutional care for longer. For children, robots have been shown to be useful in applications for those with special needs.

The natural engagement value of robots makes them a great motivational tool for getting children interested in science and engineering, or for facilitating social interaction with older people. But we should have concerns about the potential infringement of the rights of these vulnerable groups: the very young and the elderly. Society has a duty of care and a moral responsibility to do its best to ensure the emotional and psychological well-being of all its citizens regardless of their age. Looking into the future, we will see many robots *assisting* in the tasks of care. But we must be careful to leave the *practice* of care to the humans.

Robots in armed conflict and policing

It is hardly surprising that the military forces of the world have been eyeing up robots to assist them in armed conflict. Among the good uses are bomb disposal, retrieving wounded soldiers from the battlefield, and detailed aerial surveillance at the push of a button.

But there is also a massive international effort to develop robot weapons for killing people without the need for human supervision. According to the US Department of Defense, autonomous weapons systems, once launched, can select targets and engage them (apply violent force) without further human intervention. China, Russia, the US, and Israel are racing to develop robot tanks, fighter jets, submarines, and ships, with plans to use them in coordinated swarms.

Such weapons challenge international humanitarian law (the laws of war such as the Geneva Convention) and are a major threat to international security. Leaving aside the rights and wrongs of warfare, many would argue that it is morally wrong to delegate the decision to kill a human to a machine. For the past few years, I have been at the forefront of a large international coalition of NGOs and Nobel laureates who have been campaigning at the UN, and making good headway, to get a new international treaty to prohibit their development and use.

But a new international treaty will apply only to armed conflict and not to civilian policing. Police forces have been using robots for more than a decade to assist in bomb disposal, hostage release, shootouts, surveillance, and intelligence gathering. This all seems quite reasonable, but human rights advocates have been disquieted by the increasing trend to arm robots with so-called "less than lethal weapons." Desert Wolf, a South African company, has since 2014 been selling Skunk drones armed with four paintball guns that can fire eighty pepper balls per second. Now increasing demand has pushed the company to open manufacturing plants in Oman and

Brazil. Armed civil drones are spreading quickly. And in the US, North Dakota recently passed Bill 1328, allowing police to arm their quadcopters with so-called "less than lethal weapons" such as Taser stun guns (which can, in some circumstances, become lethal).

Other US police forces have been improvising robot weapons for years, and the first kill was a suspected sniper in Dallas in July 2016. There was a clear justification for this, and legal experts have pronounced that it was lawful. But a red line may have been crossed. We need to protect our police, and policing should, as far as possible, use nonviolent means. When these prove ineffective, necessity requires the level of force to be escalated gradually and proportionately to the offense being committed. This would be an enormous challenge for autonomous robots.

Autonomous robot weapons currently being developed for armed conflict could come back to haunt the civilian world. Given the appropriate circumstances, such as terrorist threats, robots could be used in large numbers to "protect" citizens. They could become essential tools to identify and track large numbers of people. With the persistent threat of further attacks, it would be difficult to put them away again. The extensive powers that robots would give to police could easily be abused in any society—and that is not a future we should want.

Robots stealing your job

Reports abound about AI and robotics creating mass unemployment within a couple of decades. In 2013, academics forecast that 47 percent of all jobs in the US could be computerized within twenty years, and in 2014 the consultancy group Deloitte followed with studies in Britain, Switzerland, and the Netherlands showing similar results. The Banks of America, UK, and Italy have expressed grave concerns.

Is there a solution? No, but there are a number of things that might help. Some have suggested that any robots should be complementary to human skills and capabilities. So robots may be used to replace or assist portions of a job rather than replacing an overall trade or occupation. But the plummeting costs of robotics and the accelerating number of tasks that can be performed could leave many without a job.

There have been a number of suggestions about what we could do with a very high level of job displacement. Bill Gates, the founder of Microsoft, has suggested that companies pay tax for a robot replacing a human. This would feed into a universal basic income (UBI), which could take a number of different forms, such as giving everyone the same amount and then taking some away for every dollar earned. Although a number of UBI trials are taking place, no country has yet introduced a policy for UBI.

It is possible that all this robotization could lead to a dystopian future of poverty and destitution for many. Alternatively, it could lead to a utopian future where we have greater freedom to do whatever we want with our time. Only time will tell.

Robot repair of climate change damage

As the impact of climate change (see Chapter 3) kicks in, our food and water supplies are predicted to be less than enough for the planet's growing population. The good news is that robots can help mitigate and prevent some of the damage.

Robot submarines go where humans struggle—up and down the deep columns of ocean water that reveal vital clues about global warming change. At Monterey Bay Aquarium Research Institute in Moss Landing, California, marine biologists are spearheading this cutting-edge research. Their underwater vehicles, like the Paragon, are doing the grunt work of today's ocean science. And submersible robots are being used to detect and repair damage to coral reefs.

At Singapore's National Water Agency, robot swans are in operation to monitor a range of physical and biological compounds in their reservoirs. Another project, also in Singapore, involves developing swarms of tiny robot sea turtles for environmental monitoring of the ocean and reducing the impact of oil spillage. And several groups around the world, including one in Scotland, are working on submersible robot systems that can diagnose damage to coral reefs and then repair them. Known as "coralbots," such robots would work in swarms, rather like bees and ants, and would be programmed to recognize corals and distinguish them from other sea objects. They could then work to repair those corals that have been damaged by bottom-trawling fishing boats or natural disasters such as hurricanes by piecing them together, allowing them to regrow.

At Michigan State University, schools of robot fish are helping to restore the Great Lakes. They communicate wirelessly to provide resource managers with a steady flow of water-quality data. The fish carry sensors that record temperature and oxygen levels and detect pollutants.

The Scottish Association for Marine Science has been working on their own custom-built Remotely Piloted Aircraft (RPA) to survey some of the world's most dangerous and inaccessible Arctic terrains to investigate the causes of melting ice. The robot plane uses a laser range-finder and a camera to measure and photograph glaciers in the polar region. This type of unique data collection by glacier experts is at the cutting edge of global warming research.

Robots play a key role in detecting chemical leaks, such as oil spills, leakages of explosive methane gas in cracked pipelines, and toxic chemicals escaping from supply pipes, which are a major hazard to humans and the ecosystem. Then there is the far from small matter of leaking water pipelines, where up to 25 percent of the world's precious drinking water is being lost. In the UAE, an aerial drone is reducing this wastage by up to 10 percent by detecting leaks.

Robot technology can make food production more efficient but less damaging to the environment where it is grown. Drones are used to test the ripeness of crops to prevent food waste. Autonomous robot farm machinery dramatically increases the efficiency of food production and is more fuel-efficient.

Environmental protection and efficient food production could be the most important tasks for robots in our future. With proper funding and international determination, these robots would become the saviors of humanity.

Our future with robots

These are just a few of the areas in which robots will impact our lives over the coming decades; there will be others that we are not even able to imagine. But it should be enough to show the enormous difficulties in charting the future of robotics. Some of the possible futures are (perhaps excessively) bleak and dystopian while others are over-optimistically wonderful utopias. Governments are just beginning to turn their attention to some of the problematic issues, and the European Parliament has recently voted on the first report on what a law on robotics needs to look like. But regardless of any new laws and regulations, it is ultimately up to all of us to pay attention to the world of robotics, both as consumers and citizens, to ensure that our future with robots is fruitful, engaging, and for the benefit of all.

THE FAR FUTURE

Time travel, the apocalypse, and living in space

16

Interstellar travel and colonizing the solar system

Louisa Preston

Since the year 2000, humanity has been living in space, albeit in a colony never greater than ten people. The International Space Station (ISS) is our first extraterrestrial outpost, an inhabited satellite orbiting the Earth, and yet no terrestrial or land-based space colonies have been built as of 2017. Nevertheless, the idea that humans will eventually travel to and inhabit other parts of our galaxy is not only a staple of science fiction, but has become part of an image of humanity's destiny, even developing into a self-imposed measure of our future success. Expanding our knowledge of the universe and our place within it is perhaps the most important reason for humanity to step into the cosmos, but it is not the only reason. There may come a time when the Earth is unable to sustain our growing numbers. To survive as a species—and protect the other life-forms we share our planet with—it will become a necessity. Space is also full of useful materials and energy, and offers a virtually endless supply of resources for us to use. But before we can take advantage of these, a lot of cultural and technological progress will be needed.

Thrusters engaged

The more we understand, or at least begin to appreciate, the sheer size of the universe, the more we realize how much we will need to progress as a species. We cannot change the distances we would need to travel to reach even the nearest stars—our neighboring star, Alpha

Centauri, is 4.3 light-years, or twenty-five trillion miles, away. If humanity's currently fastest-moving spacecraft—*Voyager*, traveling at eleven miles a second—undertook this journey, it would take over seventy thousand years. So, in the words of *Star Trek*'s Lieutenant Scott, "We cannae do it, Cap'n—we don't have the power!" One of our first goals, therefore, is to design and refine technologies, both in theory and in practice, to propel us farther and faster into the cosmos. So where should we start?

Our "old faithful" mode of interstellar transportation has been with chemical rockets using the combustion of propellants to produce exhaust gases that are accelerated through a rocket nozzle to create thrust. Currently, we can't improve upon these to help in our space adventures, as we have already maximized the energy held in their chemical bonds. However, there is still value in using them for generating fuel at a destination planet—although getting to that planet in a reasonable journey time remains a problem. We also currently rely on *electrothermal engines*, but the thrust they generate is extremely weak. They have been used since the 1970s to orient satellites but would not be our engine of choice for exploring the galaxy. In the future, we may see *ion drives* and *solar sails* propelling us through space. An ion drive is a thruster in which molecules of an unreactive fuel are given a positive or negative charge, i.e., ionized, and accelerated by an electric field to be shot out the back of the ship. The thrust is initially very low, but over a long-range mission it can deliver ten times as much thrust per pound of fuel as a chemical rocket, and could get us to Mars in thirty-nine days instead of the current six to eight months. Ion thrusters are currently used by the Dawn space probe in orbit around the dwarf planet Ceres, and were used to help it become the first spacecraft to enter and leave the orbits of multiple celestial bodies. Solar sails, on the other hand, utilize the light streaming from the sun to generate what is known as radiation pressure, which can provide significant thrust. A spacecraft with a large enough sail could eventually reach incredible speeds without carrying any fuel, but as it travels away from our solar system, and the sunlight inevitably becomes weaker, the available thrust decreases. Therefore, some form of additional fuel would be needed to keep the ship on course and indeed to eventually slow it down.

Other potential options to help us become an intergalactic species include plasma propulsion engines (essentially high-octane ion drives), thermal fission, continuous or pulsed fusion engines—which effectively attempt to re-create the power of the sun—and, at the very end of the feasibility spectrum, antimatter drives. Antimatter is matter composed of subatomic particles whose electrical charges are reversed compared with those within normal matter. This means that when it comes into contact with normal matter, both mutually annihilate each other in a burst of pure energy. This gives the highest energy density of any known process, so if antimatter is used as fuel it could provide by far the most efficient propulsion system for interstellar travel. In 2006, the NASA Institute for Advanced Concepts (NIAC) funded a team to look into the feasibility of an antimatter-powered spaceship. They calculated that just ten thousandths of a gram of antimatter would be sufficient to send a ship to Mars in forty-five days. The problem lies, however, in generating enough of this fuel, as the total amount of antimatter created in all the world's laboratories to date contains just enough energy to boil a cup of tea.

A home away from home

The faster and more powerful the ship we build, the better become our chances of reaching the stars. The ultimate goal is a spacecraft that could carry a crew through space at speeds approaching the speed of light, which presents two clear advantages. The first, perhaps obviously, is that the journey requires less time—it would take only a few years to reach the nearest stars. Secondly, if you travel at 99.5 percent the speed of light, time itself slows down, according to Einstein's theory of relativity, by a factor of ten—an astronaut would age only ten years during an interstellar journey spanning one hundred light-years. The downside to this is that they would return to an Earth where everyone had aged by one hundred years. A craft such as this would need an incredibly powerful propulsion engine as well as a great deal of shielding to protect both the ship and its occupants from high-speed collisions with asteroids or space debris. Alternatively, instead of pushing to go faster, perhaps we might slow down and embrace the time it might take to make the journey. An interstellar ark that travels at much slower speeds, say 0.2 percent the speed of light, could reach a few dozen stars in ten thousand years, yet this would require the commitment of generations of humans living aboard a Generational ship.

A *generation ship* would carry many thousands of inhabitants—generation upon generation of space farers—and be durable enough to survive the rigors of space over millennia. Such a ship would have to be entirely self-sustaining, providing energy, food, air, and water for everyone on board, and have extraordinarily reliable systems that could be maintained exclusively by the ship's inhabitants over long periods of time. The crew would essentially be families, with children born on board trained to replace their parents and grandparents in maintaining and piloting the ship as the former get old and die. These arks would have to deal with major biological, social, and moral quandaries, especially in relation to self-worth, government, and purpose—the original pioneering generations would not live to see their mission accomplished and the intermediate generations would be destined to live and die in transit without seeing tangible results for their efforts. How might this "middle child" generation feel about and cope with their forced existence on such a ship?

Perhaps a better option to combat this issue would be to put most or all of the crew into hibernation or suspended animation aboard *sleeper ships*—that way all inhabitants would survive to the end of the journey. We could also design an *embryo-carrying interstellar ship* (EIS), which would transport human embryos or DNA in a frozen or dormant state to the destination world. Yet there are obvious issues with this method, in terms of birthing and raising these individuals, so ships allowing for trips with durations comparable to a human lifetime would be preferable. The theoretical physicist Freeman Dyson conceptualized such a vessel, a *Project Orion-ship*, propelled by nuclear explosions, powered by nuclear fusion or fission. Finally, we could construct ships driven by laser power beamed from the solar system. Stephen Hawking, Yuri Milner, and Mark Zuckerberg are championing such a concept called Breakthrough Starshot—a project to build the prototype for a tiny, light-propelled robotic spacecraft capable of interstellar travel with the ability to make the journey to Alpha Centauri. They are planned to be ready for launch in under twenty years. The Starshot spacecraft will consist of a wafer-sized chip attached to a super-thin sail. This paired duo will be launched to space aboard a mothership and then propelled to the stars at up to 20 percent the speed of light by lasers beamed from a high-altitude facility on Earth. At such speeds it would take them a further twenty years to reach Alpha Centauri.

Perhaps waiting for our technology to catch up with our cosmic desires isn't the only option. What if science fiction has already found the answer? We could follow the space travel ethos of movies such as *Contact* and *Interstellar* and "simply" find a wormhole. However unlikely it might seem, we cannot entirely discount the theoretical possibility of skipping throughout the galaxy at a whim via wormhole conduits. However, we currently have no proof that this method would actually work, should we ever indeed discover a wormhole, or be remotely safe for squishy fragile life-forms such as ourselves.

Human frailty

Another great challenge facing human space exploration, alongside the huge cost and the required technological development and scientific progress, is the fragility of the human body to withstand it. The impact of long-term space travel on our delicate physiology must be properly understood if future space travelers are to survive for a prolonged or even indefinite period in space. To help us in this endeavor, the ability of technologies to shield the human body from many of the dangers of space also need to be tried and tested. Our current guinea pigs are the astronauts aboard the ISS.

Humans require an artificial life-support system to provide air, water, and food and to maintain comfortable temperatures and pressures within their spaceships, and in turn need those ships to provide shelter and protection against hazardous solar and cosmic radiation and incoming micrometeorites and space debris. Despite science fiction's depictions of space stations with rigid rotating structures creating centripetal forces to produce artificial gravity, these have not yet been built or flown, mainly because of the large size of spacecraft required. Therefore, today, the lack of gravity in space is something we need to address, as we cannot currently avoid or protect against it. The human body manages to adapt remarkably well to living in weightlessness; however, the longer it spends in space, the greater the enduring impact the lack of gravity has. Microgravity means that the body no longer needs to work against the attraction of the Earth, so it relaxes, causing effects such as muscle loss and the deterioration of bone mass as calcium oozes out of bones and leaves the body within urine. This weakens human bones over time and simulates accelerated osteoporosis. While floating around in space sounds relaxing and extremely fun, a space traveler would literally waste away if that were all he or she did. Astronauts have also found their spine to elongate while living in space—they can essentially grow an inch taller—and have also developed a swollen moon-face as the body's fluids move upwards, creating eye problems. These conditions are thankfully mostly reversible once back on solid ground and in Earth's gravity—but what if that isn't an option during a long, even generational, voyage, or on a future planetary home with a lower gravitational pull?

Finally, but no less importantly, life in the cosmos also means dealing with a very distinct lack of personal space. The best-known challenges encountered are long-term isolation, monotony, limited mobility, disrupted sleep patterns, reduced personal hygiene, and living in extremely close quarters with the same people. Privacy is indeed a luxury. The ISS, the only example of an extraterrestrial outpost, is about the size of a five- or six-bedroom house, but even so, being housebound for six months or longer is hard to cope with both mentally and physically. The assumption and hope would be that any long-duration space mission would involve a considerably larger ship than the ISS. Life in space means living with risk on a daily basis and has the potential to cause depression, insomnia, anxiety, interpersonal conflicts, and even psychosis, yet astronauts are well prepared to deal with these threats, as would be any long-duration crew. Finally, it's no secret that pre-packaged, dehydrated space food is bland, space ice cream doesn't compare to the real thing, and the extent of seasoning involves pepper suspended in olive oil to prevent it scattering around the station. Eating the same plain foods day in, day out can cause menu fatigue, a common affliction of those having to live on limited rations. Gastronomically bored astronauts are at risk of losing their appetite, leading them to consume fewer calories and ultimately lose weight and become malnourished. Any long-term voyage will need a huge repertoire of fresh tasty meal choices.

Extraterrestrial outposts

At this point let us stop and appreciate just how far humanity has come despite all the above prophecies of what might lie ahead. Since Yuri Gagarin first journeyed into outer space in 1961, we have sent humans to the Moon and robotic explorers to Venus and Mars; examined the largest asteroids; taken close-up images of Jupiter and its giant moons; flown through Saturn's rings and the icy jets emanating from one of its satellites, Enceladus; taken detailed photographs of Uranus and Neptune; finally exposed the true frozen beauty of Pluto; and even bounced onto a moving comet. Our preliminary explorations of the solar system have prepared us to now start seriously considering establishing an extraterrestrial self-sustaining colony. Yet at our current level of technology, the building of a settlement on any world other than the Earth presents a huge set of challenges—ones we need to surmount if we hope to keep hundreds or even thousands of people alive in an environment that would be undoubtedly hostile to human life.

THE MOON

For now, our most realistic colonization destinations are those worlds closest to us. Our first port of call may well be the Moon—an ideal staging post to accumulate materials, equipment, and personnel outside the confines of Earth's gravitational pull, and a useful test bed for the technologies needed to place humans on other worlds. A lunar base could increase our ability to send missions onward to Mars or into deep space and even support a bustling space tourism business. To build a habitat on the Moon is no easy feat, however, especially in gravity that is just one sixth of the Earth's. Attention must also be paid to building materials: How will they respond to the Moon's vacuum, how can they be reinforced to withstand impacts by micrometeorites traveling at up to six miles per second, and how will they react to the extreme temperature variations between day (250°F) and night (down to –243°F)?

These lunar habitats, as with any extraterrestrial abode, will be a lifeline for future Lunarians and as such must provide breathable pressurized air, water, an environment in which to grow food, protection from the harsh radiation of the sun, as well as light, warmth, and power during the long nights, each of which lasts for two Earth weeks. In 2009, more than forty permanently darkened craters near the Moon's north pole were discovered by India's probe Chandrayaan-1, containing an estimated 6.6 million tons of water ice—a resource which could be mined by a colony. A self-sustaining habitat would recycle over 90 percent of this water, producing carbon dioxide (CO2) to be pumped into a greenhouse for use by plants, which would in turn produce oxygen as a waste gas during photosynthesis, to be recycled back into the habitat.

MARS

Although the Moon is closer, it is Mars that has seemingly captured humanity's imagination as a future human outpost. Even the entrepreneur Elon Musk has plans to get one million people to Mars in the next fifty to one hundred years using SpaceX's Interplanetary Transport System (ITS)—a system of spaceflight infrastructure and technologies designed for human exploration and colonization. Terrestrial robots have been scouring Mars for the last forty years to teach us not only about our own history but what awaits us on its surface. Mars today, despite its reputation, has the most clement and almost welcoming environment in the solar system after the

Earth. If we assume we can surmount the physical obstacles of humans reaching the Red Planet, including the up to three-hundred-day journey time, radiation risk, extended periods in microgravity, and extraordinarily dangerous landing (the current success rate for unmanned craft setting down safely and in one piece on Mars is less than 30 percent), building an outpost is a reasonable goal. And once there, we would have far better conditions to work with than are present on the Moon.

Mars has a similar length of day and is tilted on its axis at an angle comparable to that of the Earth, creating similar seasons; and it has an atmosphere (albeit thin), water ice, and habitable environments. It is this climate, however, that is the main challenge we will need to overcome. Mars's atmosphere is 95 percent carbon dioxide. This means it is toxic to humans and encourages low atmospheric pressures (0.006 atm), making our existence on the surface unaided impossible. Additionally, it has only 38 percent of the Earth's gravity, is always cold (–121°F to 23°F), and there are no liquid bodies of water on its surface. Where would we live on Mars? Well, alongside the stereotypical inflatable domed housing on the surface, there are also impact craters and lava tubes across Mars that could be used to structure habitats. Indeed, these would enable larger structures to be assembled for long-term living (the only real way to live on Mars) and would provide decent protection from the outside environment. Such structures could be built in stages during a series of missions, taking inspiration from the piece-by-piece construction of the ISS. These habitats would need to be self-sustaining from day one, growing their own food, extracting water, and producing oxygen. The first Martians will therefore be of two species—plant and human—the perfect traveling companions, exchanging carbon dioxide and oxygen, keeping each other alive.

VENUS AND ICY MOONS

We cannot ignore some other interesting planetary players, despite their obvious barriers to colonization. Moving to Venus is far beyond our technical and physiological capabilities today; our fragile human bodies would die in less than ten seconds on the surface—instantly crushed by ninety-two bars of atmospheric pressure while being simultaneously cremated by 869°F of heat, with the last breath taken being of toxic gas. Life on Venus would be hot, poisonous, brutal, and short. The type of human habitat that could use the strengths of Venus's extreme conditions, however, is one that floats high in the dense Venusian atmosphere, maintained with gases of the same composition as Earth's atmosphere. The atmospheric pressure at thirty-one miles above the surface of Venus is like that on Earth at sea level (1 bar), and temperatures are just over 32°F at that altitude. If such a habitat were possible, a human could venture outside with just an oxygen mask and look out over the Venusian clouds below.

Two of the moons orbiting the gas giants Jupiter and Saturn are also of theoretical interest. If human explorers were to set up a base on Jupiter's moon Europa, the cold (down to –364°F), icy surface would actually be quite suitable (we could approach it as we do living in Antarctica), but a serious threat to life would be Jupiter's magnetosphere, which bombards Europa with deadly radiation. The best location for a base would therefore be either below Europa's icy crust or on the hemisphere of the moon that faces away from Jupiter, as this receives the least amount of radiation. On Saturn's moon Titan, colonists wouldn't need a pressurized spacesuit, just a tank of oxygen and some pretty warm clothing. With its thick atmosphere, standing on the surface of Titan would feel rather like being submerged in a swimming pool on Earth. The landscape resembles that of the Earth, and there are flat areas of land for a colony to build upon. Excitingly, Titan offers a lot to work with, as it already possesses an abundance of all the elements necessary to support life, including water ice.

Conclusions

It's difficult to consider what might be the medium- to long-term developments in space exploration using our current level of technological sophistication as the starting premise. At some point, we have to separate the possible from the probable, the realistically foreseeable from the fantasy, and the science fact from the science fiction. We know that better propulsion solutions are needed to overcome obstacles presented by the vast distances involved. The creation and selection of technological solutions to deliver the mission and literally to be the vehicle for change, as well as the development and deployment of robots to investigate potential suitable locations, present serious funding and commitment issues into R&D, lasting many years. Selection, not of people (that comes later) but of potential target "worlds," is an issue; as even a simple engineering project on our own planet involves preliminary investigations, site surveys, environmental sampling, and analysis before moving forward.

What we do know is that the human exploration of space is both dangerous and expensive. We know that humans are fragile, vulnerable, fussy about their environment, require food, water, and oxygen, have a low tolerance for the space environment, and are averse to risking their lives. Our faithful robotic explorers, on the other hand, are far less picky and can last for decades or more with comparatively minimal protection and support. It is no surprise that we have favored these loyal machines to do our exploration for us, especially as (thanks to us) they become progressively more competent and self-directed. Robots have taught us so much about our moon and Mars, highlighting the problems for us to visit in person and giving insights into the solutions. Yet humans in space provide operational flexibility, inspiration, and native intelligence—our curiosity cannot (at the moment) be matched by any machine. There are hundreds, most likely thousands, of possible habitable environments on worlds throughout our solar system and beyond, waiting for humanity to visit them. Hopefully this desire to explore new worlds will not only drive forward space exploration but help us figure out how to live successfully for the long term on our home planet as well.

17

Apocalypse

Lewis Dartnell

Everything not saved will be lost.

NINTENDO SUPER MARIO GALAXY
"QUIT SCREEN" MESSAGE FROM THE END GAMES,
BY T. MICHAEL MARTIN

Many of the chapters in this book have looked at what's appearing on the horizon in terms of our deep scientific understanding of the universe or advanced technologies that could revolutionize how we live our lives. But what I want to explore is some of the ways in which this might not go according to plan. What if the future doesn't arrive as scheduled?

In order to understand what may happen in the future we can turn to the past and learn from our own history. And, over the thousands of years since humanity developed agriculture and began settling into burgeoning towns and cities, numerous civilizations have collapsed and disappeared. In fact, the sustained development and technological progress of our own civilization over many centuries is something of an anomaly in history. So what cataclysmic events could potentially trigger an apocalypse in our own future, and crucially, what might we do today to preserve the seed of our modern knowledge to allow survivors to reboot civilization again as quickly as possible?

The collapse of any great civilization often results in a discontinuity in history, where knowledge becomes lost to the mists of time and progression stumbles and falters for a period. The most commonly referred to example is that of the "Dark Ages" in the wake of the collapse of the western Roman Empire. This narrative is slightly simplistic and also distinctly Eurocentric, as learning and technological and social development continued in the Islamic and Chinese worlds, as well as within Europe itself during this supposed period of post-collapse stagnation. For this was nevertheless a time when a number of crucial advances were made—the invention of the heavy plow that revolutionized agriculture in the clay soils across northern Europe, the tower windmill, and the mechanical clock, for example. But the early Middle Ages undeniably witnessed a profound historical hiccup after the dissolution of the Roman Empire. There was also a much earlier widespread collapse of Bronze Age civilizations around the eastern Mediterranean and an ensuing Dark Age from around 1100 BC. Likewise, many other civilizations and cultures around the world have undergone a rise and fall throughout history: the Maya, Olmecs, the Rapa Nui civilization on Easter Island, and the Indus Valley civilization, to name but a few. And it would be arrogant to assume that our own current industrialized civilization is somehow immune to sudden collapse and that it will continue indefinitely. In fact, anthropologists such as Joseph Tainter argue that as societies become more complex and interlinked, like our own, they actually become more vulnerable to sudden and catastrophic collapse.

Global catastrophic risks

Civilizations collapse for a variety of reasons: from war and invasion, natural disasters, or overexploitation and degradation of their natural environment. Hazards that may threaten our own modern, worldwide civilization in the near future are known as global catastrophic risks. These range from the distinctly plausible (pandemics) to the unlikely (asteroid impact) or even the utterly impossible (zombies!). So let us take a closer look at five realistic possibilities here.

CLIMATE CHANGE

This may well be the most likely trigger for the toppling of our modern world. We know that climate change is already happening and that it is our own activities that are driving it: carbon dioxide (and methane) emissions from our industry, transportation, and agriculture. Not only will the average global temperature continue to increase, along with a rise in sea levels from thermal expansion of the water and the melting of the ice caps, and acidification of the oceans from dissolving of all that extra CO2, but there will also be severe changes in regional climates. Global warming will cause shifts in the distribution of where rain falls, leading to increased floods in some areas and droughts in others, both of which will seriously challenge productive agriculture, particularly the amount of food we can grow to feed our rapidly multiplying human population. Access to reliable sources of fresh water could soon become major flashpoints of geopolitical tension, just as much as crude oil and other valuable natural resources, and the eruption of the first "water wars" may not lie that far in our future. Making exact predictions about the rate of climate change, and local effects, is extremely difficult with such a complex system as the Earth's atmosphere, oceans, and landmasses, and all the feedback loops this involves. One particular concern is the large amount of methane gas currently trapped in permafrost and on the ocean floor—methane is around twenty-five times more potent as a greenhouse gas than carbon dioxide, and if it were to be released it could unleash a very sudden warming. The risk is that climate change could occur so rapidly that our infrastructure proves unable to adapt and modern civilization collapses.

ASTEROID AND COMET IMPACTS

Most asteroids sit in a swirling belt between Mars and Jupiter, but some have orbits that bring them much closer to us. These are the Near Earth Asteroids, and although they offer us opportunities in the future for space mining for valuable metals, some also pose a risk of collision with Earth. We believe that our sky surveys have now identified and tracked the most potentially hazardous objects, and several plans have already been sketched out for how a space mission could deflect a troublesome asteroid onto a safer orbit (using similar technologies as space mining). We'd have much less warning of a comet strike—perhaps just a matter of months. Comets normally reside in the dark outer reaches of the solar system but can be gravitationally "nudged" into new orbits around the sun that take them plunging toward the inner solar system and the Earth. This means that not only would a comet be traveling very fast when it hits the Earth but that our telescopes might fail to spot it in time to do something about its collision course. The effects of a large asteroid or comet impact would be devastating. If such bodies dropped into the ocean they would cause tsunamis to swamp surrounding seaboards, while hitting land would throw huge amounts of pulverized rock into the atmosphere and trigger widespread wildfires. A six-mile-wide impactor such as the one thought to have triggered the mass extinction sixty-five million years ago—the "dinosaur killer"—is unlikely to strike the Earth in the foreseeable future. But even a half-mile-wide asteroid or comet would devastate a large region and could destabilize the modern world enough to trigger a terminal decline.

SUPERVOLCANOES

In many ways, the effects of a huge volcanic eruption are broadly similar to those of an asteroid impact. Obviously, if you happen to be standing too close when an eruption happens, it would be the surging lava, searing ash clouds, falling rocks, or noxious fumes that would kill you. But the wider concern is the global effect that a sufficiently large eruption would have. The ash and sulfur compounds injected high into the atmosphere would encircle the Earth and partially block out the sunlight, causing a volcanic winter that would last several years. This alone could crash global agriculture. In the longer term, the sudden cooling could even trigger another ice age. We've already had a small taste of this in recent history. The eruption of Mount Tambora in Indonesia in April 1815 led to what has been dubbed the "year without summer" and caused widespread food shortages across the northern hemisphere. And some scientists argue that the Toba eruption 71,500 years ago (also in Indonesia; this is a very volcanically active region of the planet due to the plate tectonics subduction zone) may have caused the global human population to crash to as little as just a few tens of thousands of survivors. Volcanologists today are keeping a very close eye on the huge twenty-five-mile caldera in Yellowstone National Park. This is a supervolcano that last erupted 640,000 years ago, smothering pretty much all of America west of the Mississippi river in ash. If—or rather when—the Yellowstone supervolcano next erupts, it could well disrupt our civilization enough to precipitate a collapse.

CORONAL MASS EJECTIONS

Asteroid and comet impacts aren't the only extraterrestrial hazards that could potentially trigger an apocalypse. While our sun nurtures virtually all life on Earth, it also poses a hazard to our modern, technological world. Like all stars, it regularly flings out great sprays of its aura, known as coronal mass ejections, or CMEs. These billion-ton stellar burps race out through the solar system as a hot plasma bubble of ions and electrons, laced together with a magnetic field. There is a tiny chance that such a CME is lobbed directly toward the Earth, and a particularly powerful event could be catastrophic. Sensitive electronics and solar panels in space are very vulnerable to the burst of particle radiation associated with a CME, and a large event could knock out a significant fraction of our bustling fleet of satellites, which we've become utterly dependent upon for communications, Earth observation, and GPS navigation. The precise timing signals from GPS satellites have also become crucial to coordinating many processes on the ground, such as financial transactions in the modern global economy. But the effects of a CME slamming into the Earth's magnetic field could be even more deeply damaging to our way of life. Severe disturbances to the magnetic field induce huge currents in the wires of electricity distribution grids and can permanently blow out the vital transformers. The "Carrington Event" that hit Earth in 1859 caused auroras visible far from the poles, deflected compass needles, and started fires from the sparks flying off telegraph wires. And this was in an essentially pre-electrified world; our modern society would be much worse affected, while long-lived blackouts would also hinder attempts to repair the damage. A 1989 solar storm caused a massive blackout across Quebec, and as recently as 2012 a huge CME missed the Earth by a hair's breadth; if it had hit it could have crushed a large amount of our modern infrastructure.

GLOBAL PANDEMICS

Infectious diseases have been the scourge of humanity throughout history, but every now and then a particularly virulent contagion rips across a large population as a pandemic. The Black Death is thought to have originated in the steppes of Central Asia, carried by traders along the Silk Road and then with rats aboard merchant shipping to erupt into medieval Europe in 1347. Over the next few years the plague killed around a third of the population in Europe, and up to half in the worst-hit urban areas, and is estimated to have killed a total of one hundred to two hundred million people across the whole of Eurasia. In recent history, the "Spanish Flu" broke out in March 1918, at the tail end of the First World War, and in as little as six months had spread across the entire world. Around a third of the global population became infected, and with its punishingly high death rate over fifty million people perished. If a virus with a high infectivity and death rate were to appear in the modern world, the effects could be far more catastrophic. Today, most of the human race is urban, living in densely packed cities that provide ideal conditions for the rapid spread of contagions, and regular long-haul flights shuttle people between distant continents, making containment or quarantine far harder. If even a modest fraction of the world's population were to fall sick and die at the same time the infrastructure of vital services—food production and distribution, hospitals, police forces, water treatment, electricity generation—could crumble and society disintegrate.

A backed-up "save" file for civilization

The effects of some catastrophes, such as a large asteroid impact or global nuclear war, are likely to be so widespread and devastating that survivors will be severely hard-pressed to recover quickly afterward. Other possibilities, such as a coronal mass ejection, could collapse our technological infrastructure but leave the huge human population largely unscathed in the immediate aftermath. What would likely follow would be fierce competition for dwindling resources and a savage "secondary depopulation." Perhaps the best way for the world to end, at least from the point of view of the survivors trying to rebuild again afterward, would be a sudden and virulent plague. This would rapidly remove the vast majority of the population but leave all the stuff behind; the communities of survivors could scavenge what they need while they re-learn all the skills and knowledge needed to reboot a civilization for themselves.

So what might we be able to do today to safeguard our future in case of a global catastrophe and collapse of civilization? There is a large community of people around the world, and in particular in the US, who take the possibility of an apocalypse during their lifetime very seriously. Such "Preppers" or "Survivalists" stockpile vital consumables such as canned food, bottled water, and medicines, as well as weapons for self-defense. But what about the knowledge you'd need to begin making and doing everything for yourself once again after industrialized civilization has disappeared? The most crucial scientific understanding and technological know-how has taken us centuries to accumulate over history. How could we preserve this kernel of human knowledge for posterity, in case the reset button is ever pushed and "everything not saved will be lost"? How can we help the survivors avoid another Dark Age and recover as rapidly as possible to reboot civilization from scratch?

Nowadays, there is a vast trove of information available online, from encyclopedias such as Wikipedia to practical guides and how-to videos on YouTube. But these websites are all hosted on the internet, which, although originally conceived as a robust military communications network in case of a nuclear strike, would still evaporate as soon as the grid went down after an apocalypse and the servers were starved of electrical power. (Although, there is a tongue-in-cheek entry on Wikipedia for a "Terminal Event Management Policy" that explains how to rapidly print the online encyclopedia to physical media in case of an impending global catastrophe.) However, the contents of Wikipedia are not structured in any sort of logical progression like a textbook, nor do they contain much practical information.

Ideally, to help future generations in case of an apocalypse, we'd want to compile some kind of "total book," or library of volumes, that preserves the most useful information. Unlike a DVD or computer database, a book doesn't require any technology beyond your own eyes to be able to access the content of writing and images. James Lovelock, who developed the Gaia hypothesis of the whole Earth as a self-regulating system, wrote an article in 1998 lamenting the fact that "we have no permanent ubiquitous record of our civilization from which to restore it should it fail." He described his conceptions for a "Book for All Seasons," like a complete school science textbook furnished with practical information. Kevin Kelly, a former editor of the *Whole Earth Catalog* and founder of *Wired* magazine, has also suggested a "Forever Book" or "Library of Utility": a remote mountaintop vault of perhaps ten thousand books that collectively stores the essential knowledge required to re-create the infrastructure and technology of civilization. The *Long Now Foundation* is an organization dedicated to long-term thinking; one of their projects is constructing a giant mountainside clock that is self-powered and will keep accurate time for at least ten thousand years. And unlike the other suggestions above, they have actually started gathering books for their own "Manual for Civilization" library (which I helped set up and to which I contributed a number of books, including, egotistically, my own popular science book, *The Knowledge: How to Rebuild Our World After an Apocalypse*).

And the surprising thing is that such ideas stretch back far earlier than our current generation. Encyclopedias were originally written as compendiums of all the information that an intellectual ought to remember (the word means the "circle of learning," or a well-rounded education). By the seventeenth century, however, with the exponential explosion in knowledge yielded by methodical scientific investigation, it became clear that no one person could any longer possibly retain all that was known, and the emphasis shifted to offering a summary of current knowledge for reference purposes. But encyclopedia compilers of the mid-1700s also appreciated far more acutely than we do today the fragility of even the greatest civilizations and the exquisite value of the scientific knowledge and practical skills held in the minds of the population that could be once again lost to history.

Denis Diderot explicitly regarded the role of his 1751 *Encyclopédie* as serving as a safe repository of human knowledge, to preserve it for posterity in case of a cataclysm that snuffs our civilization—just as the ancient cultures of the Egyptians, Greeks, and Romans have all been lost, leaving behind only random surviving fragments of their writing. The intention was that the encyclopedia would become a time capsule of accumulated understanding, protected against the erosion of time. This was the birth of the notion of an idealized "total book" (or at least a bookshelf of volumes) that explains systematically all that is known as well as the interrelationships between different topics. These encyclopedia compilers were also conscientious enough to include details of the experiments that demonstrate key principles, as well as diagrams of craft skills and practical know-how. In this way, such a perfect encyclopedia would present a condensed quintessence of all other written material on the planet, all arranged logically and cross-referenced, just as an omniscient being might possess in his or her memory. Such an overwhelming mass of material could never be retained at once or even comprehended by a single individual mind, but anyone could, at least in principle, educate themselves on all they might ever need to know by reading this one book.

However, although they can be read easily, paper books have their own problems. Paper rots and disintegrates if it gets damp and is also readily flammable. It would also be a major undertaking to construct enough hardened doomsday-libraries around the world to be useful for dispersed groups of survivors. But with modern technology, these seeds for rebooting civilization could be encapsulated in a much more compact format. With a Kindle or other e-reader, you can hold ten thousand books—an entire library of knowledge—in the palm of your hand. The problem here is that when the apocalypse comes and the grid goes down, you won't be able to simply plug your device into the wall to recharge. You'd experience the exasperation of having the wealth of human knowledge at your fingertips, but with no way of accessing it. So to solve this, I hacked for myself an apocalypse-proof kindle—a device loaded with all the knowledge you'd need to reboot civilization, in a ruggedized case with integrated solar panels wired into it.

You now have an entire library of practical knowledge in a portable format, and when the batteries run low you simply leave it in the sun to recharge. The screen and solar panels will eventually degrade, of course, but by that time your community should be well on the road to recovery. And with instructions saved on how to make your own paper, ink, and a rudimentary printing press, you can then download the information in the device's memory back into the lower-tech format of paper books.

A functioning society isn't just about the knowledge of how to make and do everything you require; you also need the means and tools to actually achieve this. So Marcin Jakubowski has been taking a slightly different tack—designing and building a cleverly interlinked set of machinery. When completed, the Global Village Construction Set (GVCS) will represent open-source blueprints for a complementary assemblage of fifty bits of equipment that together can provide all the infrastructure necessary for a self-sustaining community. These machines range from the simple, such as an oven, sawmill, and well-drilling rig, through to a renewable-energy steam engine (fired by biofuels) and wind turbine, to much more complex technologies such as a device to extract aluminum metal from clay, or an induction furnace for melting steel. The GVCS will support agriculture, energy generation, transportation, and manufacture. And most cleverly, the fifty items in the GVCS will be all you need to repair or fabricate any of the machines in the set. Jakubowski's stated aim is focused on helping communities in developing nations and decentralizing the means for production. But clearly the design specifications for a mutually supportive set of machinery able to provide for all the requirements of an independent society would also be exactly what is required for a recovering community in a post-apocalyptic world.

So if we do take seriously the possibility of a future global catastrophe and sudden collapse of our industrialized world, we should take steps now to preserve the kernel of all that we've achieved: the most crucial scientific knowledge and technological know-how that's taken us centuries to accumulate over history. This would be like a backed-up save file for our entire civilization, and enable the survivors to reboot a capable society for themselves as quickly as possible.

18

Teleportation and time travel

Jim Al-Khalili

I am going to be deliberately provocative in this final chapter of the book. Much of what you will have read is almost inevitably going to come to pass in the not-too-distant future—indeed much of it already has to some extent—while other more gloomy predictions describe a future that we must do our best to avoid, or at least prepare for. But what of the very distant future, possibly even beyond the time when we head off our home planet to explore and colonize the cosmos? Do ideas that are still firmly within the realms of science fiction today have any chance of being realized? There were many I could have chosen to discuss here, from telepathy to hyperdrives. But I have decided on two of my personal favorites. I am almost certain that neither will be achieved in my lifetime, but in the distant future . . . who knows?

Teleportation

The essential idea behind teleportation is the transfer of matter from one point to another without it traversing the physical space between the two points, and it has been a common theme in science fiction books, movies, and video games for longer than you might think. The earliest recorded story of a teleporter was probably Edward Page Mitchell's "The Man without a Body," written in 1877, in which a scientist invents a machine that can break down a living body into its constituent atoms, which are then sent as an electrical current through a wire to a receiver, where they are reconstituted again.

This is fascinating partly because it was written before the discovery of the electron and even before atoms were properly understood.

Jump forward half a century to 1929, when Arthur Conan Doyle published a short story called "The Disintegration Machine," about a machine that was capable of breaking up matter and reassembling it. As one of the characters in the story asks, "Can you conceive of a process by which you, an organic being, are in some way dissolved into the cosmos, and then by a subtle reversal of the conditions reassembled once more?" Two years later, the American writer Charles Fort first coined the term "teleportation" to explain the mysterious disappearance of people and objects and their supposed reappearance elsewhere. Such incidents were among the many "anomalies"—the famous Fortean phenomena—that Fort studied, along with mysterious supernatural and paranormal happenings that went beyond the boundaries of accepted scientific knowledge.

The modern idea of a teleporter was first introduced to the wider public in the 1958 American sci-fi horror movie *The Fly*, in which a scientist unwittingly mixes his DNA with that of a fly that has entered the teleportation pod with him. However, for many people around the world its most famous and enduring fictional realization has been the "transporter room" on board the Starship *Enterprise*, and the famous catchphrase it inspired: "Beam me up, Scotty." When *Star Trek* creator Gene Roddenberry first conceived of the idea in the mid-1960s, it was in fact to save money on special effects, since it was cheaper and easier to fade the characters out and then back again on the surface of a planet than have them physically fly down from the *Enterprise* in shuttlecrafts.

This is all very well of course, but what has serious science to say on the subject? The idea of transporting matter from one place to another without having to pass through the space in between may sound ridiculous but is in fact quite normal—provided you're down at the quantum level, that is. A process called quantum tunneling allows subatomic particles such as electrons to jump between two locations even when they don't have enough energy to do so. It's a bit like throwing a ball at a brick wall and having it disappear then reappear on the other side while leaving the wall intact. This is emphatically not science fiction. Indeed, the reason our sun shines and sustains all life on Earth is because hydrogen atoms are able to fuse together by quantum tunneling through what should be an impenetrable force field between them.

But an even more interesting and counterintuitive prediction of quantum mechanics, which has been verified experimentally countless times, is the idea of entanglement, which was discussed in Winfried Hensinger's chapter on quantum computing. Here, two or more spatially separated particles are connected to each other such that any measurement or disturbance made on one of them will affect its distant partner instantaneously, seemingly in violation of Einstein's theory of relativity and the speed-of-light barrier. In quantum mechanics, this is explained in terms of entangled particles being part of a single system rather than behaving as independent entities.

Consider the following analogy. Take a pair of gloves and put each in a separate box, then have one box taken to a distant location while keeping the other with you. If you open your local box and find a "left" glove inside then you know immediately that the other box contains the right glove. Of course, there is nothing mysterious about this, since it is just your state of knowledge that has changed—the other box always contained the right-handed glove. But in the quantum world, the gloves are replaced by entangled particles with each one able to spin both clockwise and anticlockwise simultaneously, a quantum superposition. On opening your local box you carry out what is called a quantum measurement, forcing it to "decide" which way it is going to spin. After all, we never see particles spinning both ways at once—that would be ridiculous, right? Quantum mechanics tells us—and experiments bear this out—that such quantum superpositions are real. What's more, as soon as you open your box to check up on the particle, the particle in the second box immediately also collapses from its superposition of spinning both ways to spinning just one way: in the opposite direction from the first particle. It's as though the act of opening your box has transmitted a quantum signal instantaneously to the other particle, telling it how to behave.

This idea of superposition and entanglement leads naturally to the notion of quantum teleportation. But can this be achieved in practice? The general idea behind quantum teleportation is to have entangled particles in the two locations, then to scan the object you wish to teleport in such a way that it is possible to transport pure information to the distant location via the entangled pair.

But to teleport even a single atom requires complete knowledge about its quantum state—basically we need to know everything about it. It had originally been thought this would be impossible due to a fundamental idea called the Heisenberg Uncertainty Principle, which states that we can never scan a quantum system to gain complete information about it in order to reconstruct it somewhere else. However, quantum entanglement provides a solution to this, because some of the information is transmitted instantaneously at the quantum level. This is supplemented by the information we gain by measuring the particle and then transmitting that separately afterward. The combined information (that transmitted quantum mechanically through the entangled pair and that obtained from the scan and transmitted at the speed of light separately) is then used to reconstruct the original object from identical raw material at the other end.

In 1993, an international team of six scientists led by IBM researcher Charles Bennett showed for the first time that the state of a particle can indeed be transferred to a distant location via quantum entanglement, giving birth to the modern idea of quantum teleportation. Since then researchers have been entangling larger and larger numbers of atoms. The problem of course is that while teleportation of a few photons of light or a bunch of atoms (of special kinds of gas cooled to near absolute zero) is possible, it is far harder to use quantum entanglement to transfer the stupendous amount of information needed to describe the special arrangement of the trillions of atoms that make up a human being.

It should be noted that what would be achieved in teleportation is more than simply creating a replica of the original particle. At the quantum level at least, transferring all the information content of a particle amounts to transferring the particle itself; we don't need to physically transfer the original particle. But be aware that to teleport an object requires the destruction of the object at A before it can be reconstituted at B. Having said that, recent tentative research, referred to as particle teleportation, suggests that it may even be possible to quantum teleport the object itself.

Just be aware that we are probably centuries away from the technology imagined in *Star Trek*.

Time travel

While teleportation relies on ideas from quantum mechanics—the theory of matter at the very smallest scales—ideas having to do with time travel stem from the theory describing the universe at the very largest scales: Einstein's general theory of relativity.

General relativity gives us our current best and most accurate description of the nature of space and time, and the fact that it does not entirely rule out the possibility of time travel means we can explore the subject seriously. It tells us that matter and energy cause space and time to warp and stretch. In fact, the mathematics of general relativity allows for the possibility of very exotic space-time shapes to exist, such as black holes and wormholes. One idea, of relevance here, is what is called a *closed time-like curve.* This is a circular path through warped space-time in which time itself is bent around on itself. If you were to travel along such a path, you would experience time going forward for you as normal. However, you would eventually arrive back at the same place in space that you started from, but before you even set off, meaning that effectively you would have traveled back in time. Such time loops form the basis of most theoretical ideas of time travel.

Although considered "unphysical" by many physicists, some have been more reluctant to dismiss time loops so quickly. The first solution of Einstein's equations of general relativity that described time loops came from W. J. van Stockum in 1937. It involved the hypothetical idea of an infinitely long cylinder of very densely packed material spinning rapidly in empty space. The mathematics describing such a scenario predict that the region of space-time surrounding the cylinder would be twisted around it and could therefore contain a time loop. Unfortunately, such a cylinder could not exist physically, because it would give space-time some very strange properties that would affect the entire universe, and we know the real universe does not have those properties.

In 1949, the Austrian-American mathematician Kurt Gödel, who worked with Einstein at Princeton's Advanced Study Institute, came up with another hypothetical scenario that was entirely consistent with general relativity, but that contained time loops. Most physicists, however, both then and now, believe that the logical paradoxes of time travel to the past are enough to rule it out and that loopholes in the laws of physics that allow for the possibility of time travel will eventually be filled through a deeper understanding, possibly from a unified theory of quantum gravity, that would pull together the two most important theories in physics: quantum mechanics and general relativity. So far, such a so-called "theory of everything" has eluded us, but we're working on it.

By the 1960s and 1970s, many more theoretical models that contained time loops were discovered by several theoretical physicists who were studying the solutions of general relativity. All involved rotating bodies that twist space-time around them. The best-known of these was an idea proposed by Frank Tipler, who published a paper in 1974 in which he took van Stockum's idea of a rotating cylinder a step further. He showed that the cylinder would need to be sixty miles long, six miles wide, and made from some exotic, very dense material. It would also have to be fantastically strong and rigid so as to avoid being squashed down along its length due to the enormous gravitational strain it would be feeling, as well as strong enough to provide the stupendous centrifugal force stopping it from flinging its matter outward as it spun at a surface speed of about half that of light. Despite all this, Tipler rightly pointed out that such difficulties were all practical problems that could be resolved by a sufficiently advanced technology.

So how might we use a Tipler cylinder as a time machine? Well, the idea is that you would travel to where the cylinder is spinning in space and orbit around it a few times before returning to Earth, hopefully arriving back in the past. How far back depends on the number of orbits you made. So even though you would feel time moving forward normally while you were orbiting the cylinder, outside the warped region of space-time you were in you would be moving steadily into the past. It would be a bit like climbing *up* a spiral staircase only to find that with each full cycle you reached the floor *below* the last one.

You might think that manipulating matter on this scale to create such a device would be nigh impossible. However, it may be that naturally occurring Tipler cylinders already exist in space. If they do—and that is highly debatable—they are called cosmic string. This is material that some cosmologists argue might have been left over from the Big Bang. It can exist either in the form of closed loops or stretching across the entire universe. Its thickness would be less than the width of an atom, yet it would be so dense that just one inch of it would weigh ten million billion tons.

One man who has thought a lot about time travel is the American astrophysicist Richard Gott, who showed how two cosmic strings moving past each other at high speed and just the right angle would form a time loop around the pair.

Ultimately, when it comes to time travel, it seems that the most feasible, or should I say the least ridiculous, way would be by traveling through a wormhole. Wormholes are exotic structures in space-time that are allowed by the equations of general relativity, which give a description of them as theoretical entities. Think of wormholes as shortcuts through space-time. They link two different regions of space together via a route that is in a different dimension from those of our own universe. And because space and time are intimately connected, the two ends of the wormhole can also in principle link two different times—one being in the past of the other. Therefore, passing through a wormhole would amount to time-traveling into the future or the past, depending on which direction you go.

Unlike black holes, for which we have ample observational evidence now, wormholes remain theoretical curiosities. Nevertheless, it may be possible one day to build a wormhole. Of course, this is not a job for twenty-first-century technology. In fact, it may never be possible. But allow me to speculate. Down at the very smallest length scales, trillions of times smaller even than atoms, is what is known as the Planck scale, where the very concepts of space and time lose their meaning and where frothy quantum uncertainty rules. Down here, all known laws of physics break down and all possible shapes and distortions of space-time will pop in and out of existence in a random and chaotic dance. Terms such as "quantum fluctuations" and "quantum foam," which are used to describe this frenzied activity, certainly do not do it justice. Within this foam, microscopic wormholes can appear and vanish very quickly, and the trick would be to somehow capture one of them and pump it up to many times its original size before it has a chance to disappear again.

So, what are we to believe? Can wormholes ever be built? Can they form time machines? Can closed time loops be formed at all in our universe and allow us to time-travel into the past? The truth is that we simply do not know for sure yet. But in the spirit of optimism I offer a quote from Frank Tipler, the physicist who published the first serious paper on how to build a time machine and who was himself borrowing a quote from the astronomer Simon Newcomb, who had written a number of papers at the turn of the century maintaining the impossibility of heavier-than-air flying machines:

> The demonstration that no possible combination of known substances, known forms of machinery, and known forms of force can be united in a practicable machine by which men shall [travel back in time], seems to the writer as complete as it is possible for the demonstration of any physical fact to be.

Well, the Wright brothers soon proved Newcomb wrong about heavier-than-air flying machines. I wonder whether one day we will see the same happen with time travel. Although I would be unwilling to bet much money on a time machine ever being built, my philosophy is that if our best scientific theories do not rule it out completely, it is important, and fun of course, to imagine how it might work.

I will end this chapter with an exciting idea that a number of theoretical physicists are now thinking seriously about. It may—just may—be that teleportation and time travel are intimately connected. A new idea, known among physicists as ER=EPR, suggests there may be a deep and profound link between quantum entanglement (the teleportation idea) and wormholes (the time-travel idea). Two papers, published by Einstein and his collaborators in 1935, which had hitherto been thought of as completely unrelated, may turn out to describe the same concept. The EPR paper (from the initials of its three authors, Einstein, Podolsky, and Rosen) was the first to describe the weirdness of quantum entanglement in the way two distant particles are connected instantaneously, an idea that Einstein in particular felt was impossible and that therefore hinted at our incomplete understanding of quantum theory. The second paper (the ER bit, made up of the Einstein and Rosen subset of the team) was the first work to describe the idea of a wormhole, known then as an Einstein-Rosen bridge.

So here we are, over eighty years after these two papers were published, asking a daring question: What if pairs of entangled particles are in fact able to "communicate" with each other *because* they are joined by a wormhole. The more I read up on and think about this crazy idea, the more I like it. It's just so neat. Thus, wormholes, assuming they are physically possible of course, could act as both teleporters and time machines. Wouldn't that be the coolest thing ever?

For now, of course, the subject of this chapter sits quite firmly in the realms of science fiction—and of course in the mathematical equations of the more daring theoretical physicists.

Whatever the outcome, I am utterly convinced that there will be plenty of surprises in science in the decades, and centuries, to come.

So, let's use our new knowledge wisely.

FURTHER READING

THE FUTURE OF OUR PLANET

Adventures in the Anthropocene: A Journey to the Heart of the Planet We Made, Gaia Vince, Milkweed Editions, 2015.

Climate Change (What Everyone Needs to Know), Joseph Romm, Oxford University Press, 2015.

The Future: Six Drivers of Global Change, Al Gore, Random House, 2013.

Homo Deus: A Brief History of Tomorrow, Yuval Noah Harari, Harper, 2017.

Population 10 Billion, Danny Dorling, Constable, 2013.

Scale: The Universal Laws of Growth, Innovation, Sustainability, and the Pace of Life in Organisms, Cities, Economies, and Companies, Geoffrey West, Penguin Press, 2017.

Smart Cities, Digital Nations: Building Smart Cities in Emerging Countries and Beyond, Casper Herzberg, Roundtree Press, 2017.

Tomorrow's World: A Look at the Demographic and Socio-Economic Structure of the World in 2032, Clint Laurent, Wiley, 2013.

THE FUTURE OF US

Citizen Cyborg: Why Democratic Societies Must Respond to the Redesigned Human of the Future, James Hughes, Basic Books, 2004.

Creation: How Science Is Reinventing Life Itself, Adam Rutherford, Penguin, 2014.

The Gene: An Intimate History, Siddhartha Mukherjee, Scribner, 2017.

Happy-People-Pills for All, Mark Walker, Wiley, 2013.

Life at the Speed of Light: From the Double Helix to the Dawn of Digital Life, J. Craig Venter, Penguin, 2014.

The Patient Will See You Now: The Future of Medicine Is in Your Hands, Eric Topol, Basic Books, 2016.

Spillover: Animal Infections and the Next Human Pandemic, David Quammen, W. W. Norton & Company, 2013.

Superintelligence: Paths, Dangers, Strategies, Nick Bostrom, Oxford University Press, 2016.

THE FUTURE ONLINE

AI: Its Nature and Future, Margaret A. Boden, Oxford University Press, 2016.

Cloud Computing (MIT Press Essential Knowledge Series), Nayan B. Ruparelia, MIT Press, 2016.

Computing with Quantum Cats: From Colossus to Qubits, John Gribbin, Prometheus Books, 2014.

The Economic Singularity: Artificial Intelligence and the Death of Capitalism, Calum Chase, Three Cs, 2016.

Enchanted Objects: Design, Human Desire and the Internet of Things, David Rose, Scribner, 2015.

The Technological Singularity, Murray Shanahan, MIT, 2015.

MAKING THE FUTURE

The Industries of the Future, Alec Ross, Simon & Schuster, 2017.

Innovation and Disruption at the Grid's Edge, Fereidoon Sioshansi (ed.), Academic Press, 2017.

Made to Measure: New Materials for the 21st Century, Philip Ball,

Princeton University Press, 1999.

Stuff Matters: Exploring the Marvelous Materials That Shape Our Man-Made World, Mark Miodownik, Mariner Books, 2015.

We Do Things Differently: The Outsiders Rebooting Our World, Mark Stevenson, The Overlook Press, 2018.

THE FAR FUTURE

Black Holes, Wormholes and Time Machines, Jim Al-Khalili, CRC Press, 2012.

Emigrating Beyond Earth: Human Adaptation and Space Colonization, Cameron M. Smith, Springer, 2012.

Global Catastrophic Risks, Nick Bostrom and Milan M. Cirkovic, Oxford University Press, 2011.

The Knowledge: How to Rebuild Civilization in the Aftermath of a Cataclysm, Lewis Dartnell, Penguin, 2015.

Packing for Mars: The Curious Science of Life in the Void, Mary Roach, W. W. Norton & Company, 2011.

Physics of the Future: How Science Will Shape Human Destiny and Our Daily Lives by the Year 2100, Michio Kaku, Anchor, 2012.

ABOUT THE AUTHORS

Jim Al-Khalili OBE is a British physicist, author, and broadcaster. He is currently Professor of Theoretical Physics and Chair in the Public Engagement in Science at the University of Surrey. As well as his writing and his research in nuclear physics and quantum biology, he has hosted many TV and radio productions about science, including the weekly BBC Radio 4 program *The Life Scientific*. He is a recipient of the Royal Society Michael Faraday Prize and was the 2016 winner of the inaugural Stephen Hawking Medal for Science Communication.

Philip Ball is a writer and author. Having trained as a chemist and physicist, he worked for many years as an editor for *Nature*. He now writes regularly on science and its interactions with the arts and the wider culture. His books include *Bright Earth: The Invention of Colour*, *The Music Instinct*, *Curiosity: How Science Became Interested in Everything* and *Invisible: The Dangerous Allure of the Unseen*. His book *Critical Mass* won the 2005 Aventis Prize for Science Books. Currently he is a presenter of BBC Radio 4's "Science Stories," and his latest book is *The Water Kingdom: A Secret History of China*.

Margaret A. Boden OBE is Research Professor of Cognitive Science at the University of Sussex, where she helped develop the world's first academic program in cognitive science. She holds degrees in medical sciences, philosophy, and psychology, and integrates these disciplines with AI in her research. She is a Fellow of the British Academy and of the Association for the Advancement of Artificial Intelligence (and its British and European equivalents). Her work has been translated into twenty languages. Her most recent book is *AI: Its Nature and Future*.

Naomi Climer is an engineer. Her career has been in the broadcast and communications technology industry, including at the BBC, ITV, and Sony in Europe and the US. Naomi is Past President of the Institution of Engineering and Technology (IET), Governor of the National Film and Television School (NFTS), Chair of the Council of the International Broadcasting Convention (IBC), and consultant to the board of Sony's UK Technology Centre.

Lewis Dartnell is an astrobiology researcher at the University of Westminster. He studies how microbial life, and signs of its existence, might persist on the surface of Mars exposed to the bombardment of cosmic radiation, and how we could detect it. Lewis features regularly on television and radio talking about science, and his past books have included *Life in the Universe: A Beginner's Guide* and *The Knowledge: How to Rebuild our World from Scratch* (www.the-knowledge.org), a *Sunday Times* Book of the Year.

Jeff Hardy is a Senior Research Fellow at the Grantham Institute – Climate Change and the Environment at Imperial College, London, where he researches what the future low-carbon energy system might look like, how people will engage with it, and what businesses will be operating in it. Previously he was Head of Sustainable Energy Futures at the UK energy regulator, Ofgem, and Head of Science for Work Group III of the Intergovernmental Panel on Climate Change. He has also worked at the UK Energy Research Centre, the Royal Society of Chemistry, the Green Chemistry Group at the University of York, and at Sellafield as research chemist in a nuclear laboratory.

Winfried K. Hensinger is a Professor of Quantum Technologies at the University of Sussex. He obtained his PhD at the University of Queensland, demonstrating novel and strange quantum effects with ultracold atoms. During his PhD research, he spent an extended period at NIST in Gaithersburg, US, in the group of Nobel laureate William Phillips. Professor Hensinger now heads the Sussex Ion Quantum Technology Group and is the director of the Sussex Centre for Quantum Technologies. He recently published the first practical blueprint for building a large-scale quantum computer, and his group works on the construction of such a device.

Adam Kucharski is an assistant professor at the London School of Hygiene & Tropical Medicine, where he works on infectious disease outbreaks. He studied at the University of Warwick before completing a PhD in mathematics at the University of Cambridge. Winner of the 2012 Wellcome Trust Science Writing Prize, Kucharski has written for the *Observer*, *New Scientist,* and *Wired*. His debut book, *The Perfect Bet: How Science and Math Are Taking the Luck Out of Gambling*, was published in 2016.

John Miles is a Fellow of Emmanuel College, Cambridge, and is the Arup/Royal Academy of Engineering Professor of Transitional Energy Strategies at the Department of Engineering. His special interests include the technology and economics of future transportation systems, with a particular emphasis on energy efficiency and environmental impact. He was the founding chairman of the UK Automotive Council's Working Group on Intelligent Mobility, where he was responsible for the production of the Council's Intelligent Mobility Technology Roadmap and several reports on Intelligent Mobility.

Anna Ploszajski is a materials scientist and engineer by day and science communicator by night. She regularly performs stand-up comedy about materials, produces a podcast called *'rial talk*, and writes pieces about the wonders of materials for publications such as *Materials World*. In 2017 Anna won Young Engineer of the Year from the Royal Academy of Engineering and made it to the UK final of FameLab. In her spare time she plays the trumpet and is training to swim the English Channel.

Aarathi Prasad is a writer with an academic background in molecular genetics. She has developed and presented science documentaries for BBC1, BBC Radio 4, Channel 4, National Geographic, and the Discovery Channel. She writes for a number of publications and is the author of *In the Bonesetter's Waiting-Room: Travels in Indian Medicine* and *Like A Virgin: How Science is Redesigning the Rules of Sex*.

Louisa Preston is a UK Space Agency Aurora Research Fellow in Astrobiology at Birkbeck, University of London. She has worked on projects for NASA and the Canadian, European, and UK space agencies studying environments across the Earth, where life is able to survive our planet's most extreme conditions, and using them as blueprints for possible extraterrestrial life-forms and habitats. She is an avid science communicator and has spoken about the search for life on Mars at the TED Conference in 2013. Her first book, *Goldilocks and the Water Bears: The Search for Life in the Universe*, is published by Bloomsbury Sigma.

Adam Rutherford is a geneticist, writer, and broadcaster. He presents BBC Radio 4's flagship science program *Inside Science* and a host of other programs on television and radio. He has also worked as scientific consultant on a number of films, including *World War Z* (2013) and Alex Garland's Oscar-winning *Ex Machina* (2015). He is the author of, among others, *A Brief History of Everyone Who*

Ever Lived: The Human Story Retold Through Our Genes.

Noel Sharkey is Emeritus Professor of AI and Robotics at the University of Sheffield, co-director of the Foundation for Responsible Robotics, Chair of the NGO: International Committee for Robot Arms Control, and head judge on BBC's *Robot Wars.* Noel has moved freely across academic disciplines from psychology to AI to linguistics and to computer science, and from machine learning to engineering and robotics, and now to the ethics of technology. He has held research and teaching positions in the US (Yale and Stanford) and the UK (Essex, Exeter, and Sheffield).

Dame **Julia Slingo** FRS is a British meteorologist and climate scientist. She was the Chief Scientist at the Met Office between 2009 and 2016. Throughout her career she has brought innovative approaches to understanding and predicting weather and climate using complex models. Her special interests are tropical weather and climate variability. Dame Julia was elected a Fellow of the Royal Society in 2015 and Foreign Member of the US National Academy of Engineering in 2016.

Gaia Vince is a writer and broadcaster specializing in science, society, and the environment. She has been an editor at *Nature Climate Change*, *Nature*, and *New Scientist.* She writes for newspapers and magazines in the UK, US, and Australia, and presents science programs for radio and television. Her first book, *Adventures in the Anthropocene: A Journey to the Heart of the Planet We Made*, won the Royal Society Winton Prize for Science Books in 2015.

Mark Walker is a Professor of Philosophy in the Department of Philosophy at New Mexico State University, where he holds the Richard L. Hedden Chair of Advanced Philosophical Studies. His first book, *Happy-People-Pills for All* (2013), argues for creating advanced pharmaceuticals to boost the happiness of the general population. His latest book, *Free Money for All* (2015), argues for an unconditional basic income of $10,000 for all US citizens.

Alan Woodward began as a physicist and became increasingly involved with computer science while doing signal processing as part of his post-graduate research. After many years working for the UK government he continued to conduct essentially the same work while in industry. Latterly, he returned to his roots in academia as a visiting professor at the University of Surrey while continuing to advise organizations such as the European Police Office. Alan is best known for his work in cybersecurity, which encompasses a wide range of topics from quantum physics to computer science.

INDEX

W